Light-Beam Communications

by

Forrest M. Mims, III

HOWARD W. SAMS & CO., INC.
THE BOBBS-MERRILL CO., INC.
INDIANAPOLIS · KANSAS CITY · NEW YORK

Contents

Preface

In today's world of vastly overcrowded communication channels, the immense information-carrying potential of light is very appealing. Although it has been nearly a century since Alexander Graham Bell demonstrated voice transmission over a light beam, until recently most light-beam communications research was military in nature. In 1960, however, the advent of the laser and efficient solid-state light sources stimulated an intense research effort which has already resulted in more than a dozen commercial optical communication systems and scores of experimental units.

Unlike radio, light-beam communicators are very difficult to jam or intercept. Also, light can carry far more information than radio, and the FCC imposes no licensing requirements on optical communicators since the narrowness of their beams prevents them from interfering with one another. Of course, optical communication through the atmosphere is impeded by the adverse effects of weather. Nevertheless, optical communications have immensely practical applications as fiber-optic links, short-range atmospheric communications, and space links.

This book begins with a fairly detailed treatment of the history of light-beam communications. The chapters on the principles of light-beam communications, light sources, detectors, and optics supply technical background. The final chapter describes many experimental and commercial optical communicators. A supplemental book now in preparation, *Light-Beam Communication Circuits,* will present more than a dozen plans for making workable light-beam communicators and will provide details on their operation.

Several individuals and corporations provided invaluable assistance in the preparation of this book. Duncan Campbell, president of American Laser Systems, Inc. supplied laser equipment, photographs, extensive technical information, and valuable discussions. Dr. Richard Glicksman of RCA supplied detailed technical information about newly developed injection lasers and also supplied photographs and sample lasers. RCA supplied an avalanche detector. Bell Telephone Laboratories supplied numerous photographs, including two early prints of Bell's original Photophone. Metrologic Instruments, Inc. provided photos and technical information about modulated HeNe lasers. Finally, my wife Minnie took time from her busy schedule to type the manuscript. To these individuals and firms I express my deepest gratitude.

Forrest M. Mims, III

This book is dedicated to the memory of
Alexander Graham Bell (1847-1922),
the father of light-beam communications.

CHAPTER 1

Introduction

Communication by light beams is not new. Since the beginning of recorded history, man has employed signal fires or sunlight reflected from mirrors for the transmission of information. These rather primitive forms of light-beam communications are still in use today. Military pilots, for example, are issued signal mirrors with which they can attempt to attract the attention of searchers should their aircraft go down in a remote area. Every New Year's Eve, weather permitting, citizens of Colorado Springs, Colorado exchange mirror signals with men climbing nearby Pikes Peak in preparation for an annual fireworks display from the summit of the snow-covered mountain.

Prior to 1880 and the introduction of the voice-modulated light beam, the most significant advance in light-beam communications was the development of the signal light. This device consisted of a lantern blocked by a movable grid. The grid could be opened and closed to permit the transmission of coded information. Signal lights such as this found widespread use by ships at sea and are still used today by military vessels when radio silence is in effect.

VOICE-MODULATED LIGHT-BEAM COMMUNICATIONS

Credit for achieving the first wireless transmission of encoded signals generally goes to Guglielmo Marconi, the celebrated Italian inventor who changed radio from a laboratory curiosity into an ocean-spanning practical reality. Of course, it is impossible to downgrade the significance and eventual impact of Marconi's work, but coded signals had been transmitted for thousands of years before Marconi by using sunlight reflected from mirrors during the day and by using lanterns during the night.

Similarly, credit for the first wireless transmissions of the human voice generally goes to A. Frederick Collins who in 1899 sent his voice over a distance of several blocks using a radio transmitter in Naberth, Pennsylvania. It is a little known fact that Alexander Graham Bell and his laboratory assistant, Sumner Tainter, were transmitting the human voice over beams of reflected sunlight in 1880, nearly two decades before Collins' pioneering voice transmissions via radio.

ALEXANDER GRAHAM BELL'S PHOTOPHONE

One of the most significant developments in the long history of light-beam communications occurred in 1880. Seven years earlier it had been accidentally discovered that selenium bars used as multimegohm resistors in a system designed to test submerged transoceanic cables were affected by sunlight. Willoughby Smith had designed the apparatus, and his assistant, Mr. May, was the first to observe that the resistance of the selenium bars was reduced in the presence of light.

Smith was fascinated by May's discovery and soon confirmed the observation with a carefully designed experiment. Other investigators also became excited about the new discovery and authored technical papers on their experiments and findings.

How Bell came upon the idea of transmitting speech over a light beam is not known. However, an interesting pair of articles in the April 25, 1878 edition of the scholarly British journal *Nature* may shed some light on where he obtained his inspiration. On pages 512-513 of the journal is a brief paper by Robert Sabine entitled "Action of Light on a Selenium (Galvanic) Element."

Sabine's paper showed how the light sensitivity of selenium could be utilized by forming one of the battery's electrodes from selenium. Using this technique, he found ". . . that the slightest shadow or other variation in the intensity of the light caused a considerable variation in the electromotive force of the couple and a consequent indication." More precisely, Sabine found that the presence of light on the battery increased its output by 0.15 volt.

Sabine's paper was immediately preceded by an article by Professor W. F. Barrett on "Early Electric Telephony" which detailed several historical precedents to Bell's first telephone. Barrett praised the early workers for their ingenuity but concluded by recognizing the primacy of ". . . the sensitive and beautiful instrument discovered by Prof. Bell. . . ."

Bell almost certainly read Barrett's paper. The intense controversy over the validity of his 1875 telephone patent and the resultant court battles with Western Union made him acutely sensitive to the very subject reviewed in such detail by Barrett. Furthermore, Bell and his bride were in England for their honeymoon when the issue of *Nature* carrying the two papers appeared.

That Bell also saw and probably read Sabine's paper is almost equally certain. The first part of the relatively brief paper was squeezed between the end of Barrett's article, which ended with the words ". . . invented and patented by Prof. Graham Bell," and a footnote by Barrett which also mentioned Bell by name.

A rereading of Bell's speech before the Royal Institution of Great Britain on May 17, 1878, only three weeks after the cover date of the *Nature* edition, provides even more-convincing evidence; for in this talk, Bell made his first public suggestion regarding the use of a variable light to produce sound: "If you insert selenium in the telephone battery and throw light upon it, you change its resistance and vary the strength of the current you have sent to the telephone, so that you can hear a shadow."

In an 1880 address before the American Association for the Advancement of Science, Bell casually mentioned Sabine's paper but provided no hint as to when he first read it. Since the Royal Institution address specifically refers to using selenium in a battery and uses the word "shadow," both mentioned by Sabine, and since Sabine's paper appeared so shortly before the speech, it is almost certain Bell read the paper *before* his May 17th speech.

All this gives a fascinating historical insight into how an inventive genius like Bell caught the spark of an idea and added the ingredient necessary to create a brilliant invention. Sabine used a meter to see the effect of light on the miraculous selenium; Bell added a telephone receiver, his own invention, and was able to both see and hear what Sabine had only been able to see.

Upon his return to the United States, Bell followed up his idea with some actual experiments. But, since the resistance of the selenium was too high, the experiments were unsuccessful. In *Alexander Graham Bell and the Conquest of Solitude*

(Little, Brown, and Co., Boston, 1973), perhaps the best biography about the famous scientist, author Robert V. Bruce provides a detailed and fascinating account of what happened next.

Frequently quoting from Bell's notes and correspondence, Bruce describes how Bell hired a young lab assistant, Sumner Tainter, to assist him in much the same way that Tom Watson had several years earlier on the invention of the telephone. On January 22, 1880, the two researchers made a formal statement in their laboratory notebook of their intention to perfect a means of transmitting voice over a light beam. On February 19, 1880, less than a month later, Bell wrote in his notebook that he and Tainter had finally solved the problem of voice modulating a light beam.

A week later in a letter to his father, Bell displayed the enthusiasm which followed the new discovery: "I have heard articulate speech produced by sunlight! I have heard a ray of the sun laugh and cough and sing!" He went on to describe some possible applications for his invention: "Can imagination picture what the future of this invention is to be! . . . We may talk by light to any visible distance without any conducting wire. . . . In warfare the electric communications of an army could neither be cut nor tapped. On the ocean communication may be carried on . . . between . . . vessels . . . and lighthouses may be identified by the sound of their lights."

The method that Bell aind Tainter used to impress voice upon a light beam seems disarmingly simple when viewed with hindsight. Sunlight was allowed to illuminate a thin, flexible mirror. When voice was directed toward the mirror, as shown in Fig. 1-1, the undulatory nature of the sound waves caused the mirror to respond in kind. Normally, the mirror reflected a uniform beam of light, but the alternately convex and concave surface of the mirror produced by the sound waves caused the light beam to be distorted in step with the voice. The result was an intensity-modulated beam of light.

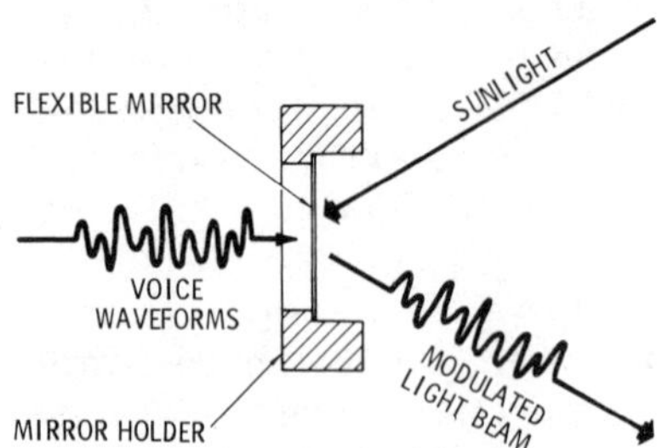

Fig. 1-1. Alexander Graham Bell's mechanical light-beam modulator (1880).

Bell and Tainter detected the modulated sunlight with a selenium cell in series with a telephone receiver and battery. The detector circuit, however, was evidently far more difficult to assemble than the transmitter. The primary problem was the very high resistance of the selenium employed in the initial detectors.

Some of the earliest selenium detectors had been perfected by Dr. Werner Siemens, and their fabrication resembles the construction of some modern solid-state components. A lattice of fine platinum or copper wires was made and placed on a perforated mica plate. A drop of molten selenium was placed on the wire lattice, and a second mica sheet was pressed over the assembly to spread the selenium over and between the wires in the lattice. The cells, which were about ten millimeters in diameter, were encapsulated in a thin layer of paraffin.

Unfortunately, the dark resistance of these cells was typically measured in the millions of ohms. Bell found that when brass was used, much better contact was made. The result was low-resistance cells (300-ohms dark resistance) which were well suited for use in a telephone-battery series circuit.

Bell and Tainter constructed two main types of selenium detectors. One consisted of a circular array of perforations in one of two brass plates separated and insulated from one another by a sheet of mica. Pins in the lower, unperforated plate protruded through the perforations in the upper plate. Molten selenium was placed in each of the perforations to give an array of circular detectors. This type of detector was designed to be used with a focusing lens.

The second detector was in the form of a cylinder to permit use with a parabolic reflector. The cylinder face was covered with a large number of metal and mica discs. The space between adjacent discs was filled with molten selenium.

Fig. 1-2 is a circuit diagram of Bell's apparatus which appeared in *Proceedings of the American Association for the Advancement of Science* (A. G. Bell, "On the Production and Reproduction of Sound by Light," Vol. 29, 1880). The diagram effectively shows the simplicity of the concept.

Fig. 1-3 is a drawing from this period, showing the light-beam transmitter system. The square object opposite the speaker is a mirror designed to collect sunlight. Another period drawing showing the light-beam receiver is reproduced as Fig. 1-4.

On March 26, 1880, Bell and Tainter obtained a range of 82 meters (269 feet) with their equipment. Encouraged by the range potential of the apparatus, they conducted a long-

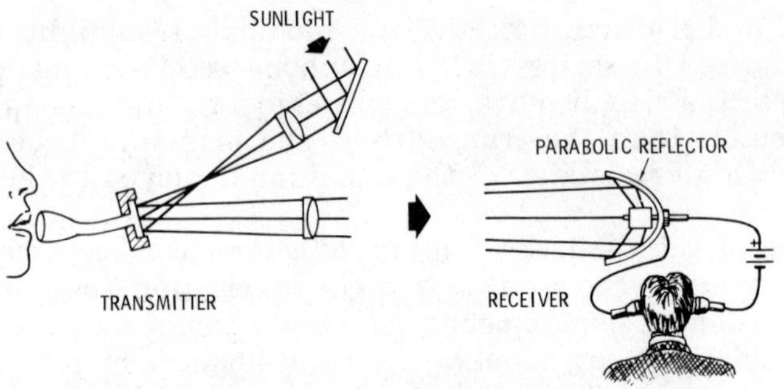

Fig. 1-2. Photophone operating diagram (1880).

range experiment on April 1st of the same year. Tainter carried the sunlight transmitter to the roof of the Franklin School in Washington, D.C. and projected a beam of reflected sunlight to a window of Bell's laboratory, 213 meters (699 feet) away. When the beam was aligned so that it illuminated the selenium detector assembly, Tainter uttered these words: "Mr. Bell, if you hear what I say, come to the window and wave your hat." In a speech several years later, Bell reported "... that I waved with vigor, and with enthusiasm which comes to a man not often in a lifetime."

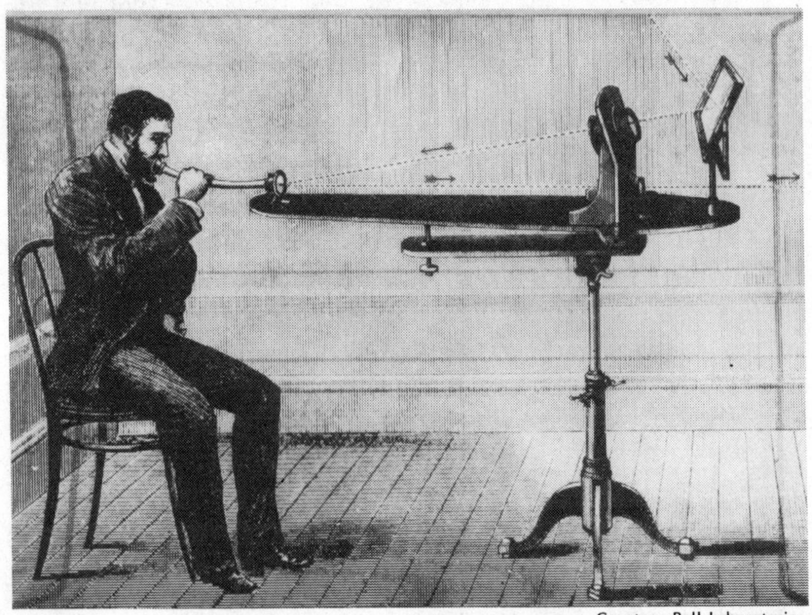

Courtesy Bell Laboratories

Fig. 1-3. The original Photophone transmitter.

Bell described his experiment with the Photophone, the name
he gave his new invention, in great detail in a paper he presented before the American Association for the Advancement
of Science in 1880. In that talk, he gave due credit to the contribution of his peers who had first discovered the light sensitivity of selenium and had made usable detectors from the
element. He also gave credit to A. C. Brown and others for
having independently conceived other methods for voice modulating a light beam, though Bell and Tainter were the first
to put the concept to practice. Bell was liberal in his acknowl-

Fig. 1-4. The Photophone receiver system.

edgements because lawsuits over the originality of his telephone invention had caused him a great deal of distress.

Bell's talk was remarkable in that he clearly anticipated
many of the basic light-modulation schemes employed since
1880, including sophisticated variable-polarization schemes
now being utilized by laser-communication engineers. Furthermore, he provided a detailed description of how intelligi-
ble information could be sent over a light beam *without* a selenium detector or even a telephone receiver. According to his
description, the beam from a transmitter could be focused
upon a thin disc of hard rubber connected to the ear by a hollow tube and a distinct musical tone could be heard when a

rotating wheel containing a circle of perforations was placed in the light beam.

Bell concluded ". . . *that sounds can be produced by the action of a variable light from substances of all kinds when in the form of thin diaphragms*" (italics added). He also found that "a musical tone can be heard by throwing the intermittent beam of light into the ear itself." These remarkable discoveries have not received wide circulation due to the superiority of electronic light detectors. However, experimenters might be able to expand upon this early work and find some practical application for it. In our energy-conscious world of today, the appeal of a communication system that uses free light from the sun and requires only an incredibly simple, nonelectronic receiver is irresistible.

THE IMPACT OF THE PHOTOPHONE

To his death, Bell regarded the Photophone as his most significant invention. Indeed, in 1921, shortly before he died, he said that the Photophone was even more important than the invention for which he is best remembered, the telephone.

Unfortunately, neither public nor scientific reception to the Photophone was as enthusiastic. Although the National Bell Telephone Company purchased rights to the invention, the firm's president, William H. Forbes, expressed some skepticism in his acceptance speech: "Whether this discovery ever approaches the telephone itself in practical importance or not, it is no less remarkable and a thing which we should be glad to possess" (*Alexander Graham Bell and the Conquest of Solitude,* p. 339).

The *New York Times,* a newspaper notorious for its mocking criticism of scientists and their inventions, was particularly harsh. In an August 30, 1880 editorial written soon after Bell's speech before the American Association for the Advancement of Science, the paper asked if there would be ". . . a line of sunbeams hung on telegraph posts?"

Despite later developments which have significantly improved upon earlier light-beam communication techniques, even biographer Robert Bruce, in his 1973 biography about Bell wrote that: "eventually the achievement of coherent light in the laser beam would greatly extend its possible range. For all that, the Photophone concept still remains one of limited utility. Nor is its scientific significance great—or even perceptible" (*Alexander Graham Bell and the Conquest of Solitude,* p. 343). As we shall soon see, advances in semiconductor technology and in the field of optoelectronics in general, give renewed hope to Bell's claims for his Photophone.

AFTER THE PHOTOPHONE

Relatively little work was done with the Photophone concept after the experiments of 1880. In 1893 Bell exhibited a non-electronic version of the device at the Columbian Exposition. The apparatus, which Bell labeled a Radiophone, incorporated a lampblack receiver and permitted visitors to send their voice over a 30-meter (100-foot) beam of light.

In 1897 AT&T advanced the concept by replacing Bell's voice-modulated mirror with an electronically modulated arc. Three years later, a German named Simon demonstrated a voice-modulated carbon arc with a range of one kilometer. Ruhmer, another German worker, obtained even greater ranges in 1904. The fast moving field of radio halted further work on the Photophone concept until World War I when the need for secure military-communication systems arose.

WORLD WAR I DEVELOPMENTS

Bell had predicted a military application for the Photophone, and World War I proved him right. Both sides developed a variety of experimental photophones, some with impressive range capability. The most important of these systems will be described in the following sections.

British Admiralty Photophone

Between February and October of 1916, Dr. A. O. Rankine of the Department of Physics in London's University College conducted experiments in light-beam communication for the Admiralty Board of Invention and Research. Rankine reported upon the results of his work in a lengthy paper published in the *Proceedings of the Physical Society of London* in 1919, the delay between the experimental study and its publication being due to military-security requirements.

Although other experimenters had begun using the carbon arc rather than the sun, as a radiant source, Rankine preferred the earlier sun approach. He correctly noted that electrically modulated carbon arcs have finite frequency-response limitations and that voice modulation would ". . . thus impose no more than a ripple of small amplitude upon the already powerful beam of light emitted from the arc." Carbon arcs were best modulated, in Rankine's view, by external means.

Rankine's work was essentially a continuation of Bell's proposal to use a movable grid assembly for the external modulation of a light beam. However, Rankine noted that Bell's proposal suffered from a lack of practicality in that the rapid mechanical movements and the mass of the grid itself would preclude a practical system. Rankine proposed substituting the *image* of a pair of grids for the actual grids.

Fig. 1-5 is a diagram of Rankine's apparatus. A 1-cm concave mirror, A, can be mechanically voice modulated, either directly or indirectly. Rankine preferred the direct technique in which the mirror was mounted to the lever that normally carried the needle in a gramophone recording machine. Speech entering the gramophone trumpet successfully modulated a mica diaphragm, the lever, and the mirror.

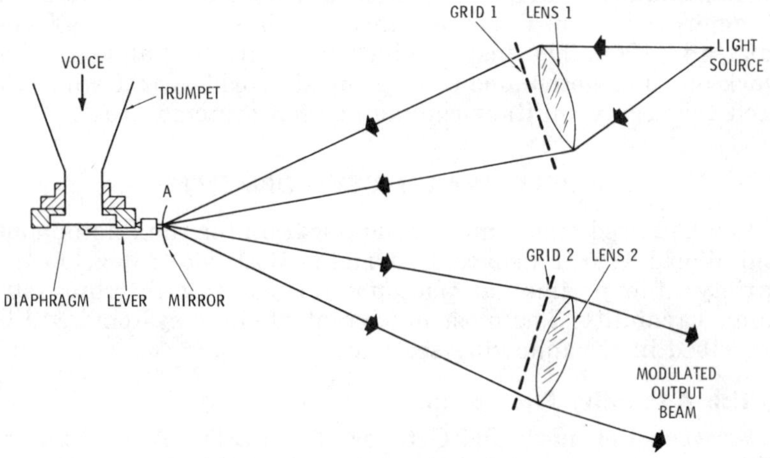

Fig. 1-5. Rankine's light-beam transmitter (1916).

The mirror is illuminated by a source whose radiation is focused by a lens which is backed by an opaque grid. The light is reflected to a second lens which is also backed by an opaque grid. With this system in proper alignment, it is clear that for one position of the mirror, the transparent spaces between the lines of the first grid will be exactly superimposed over the lines of the second grid and *no* light will emerge from the system. For another position of the mirror, the transparent spaces between the lines of each grid will be exactly superimposed and a maximum of 50 percent of the light collected by the source lens will be emitted by the system. Therefore, the light emerging from the system will have an intensity ranging from 0 to 50 percent of the light collected by the source lens.

Rankine spent considerable time perfecting his grid-modulation scheme. One persistent problem was overmodulation resulting from excessive grid-image movement. He found that the light passing through the system might be interrupted two or more times when only a single interruption was intended. Since the grid image in Rankine's system was visible to the operator, a conscious effort could be made to avoid overmodulation by watching the response of the grid to voice.

Rankine also studied an indirect modulation scheme, which had been developed earlier by Bell. In this system, Rankine affixed the movable mirror of the transmitter to a reed-type telephone receiver as shown in Fig. 1-6. Bell had used the

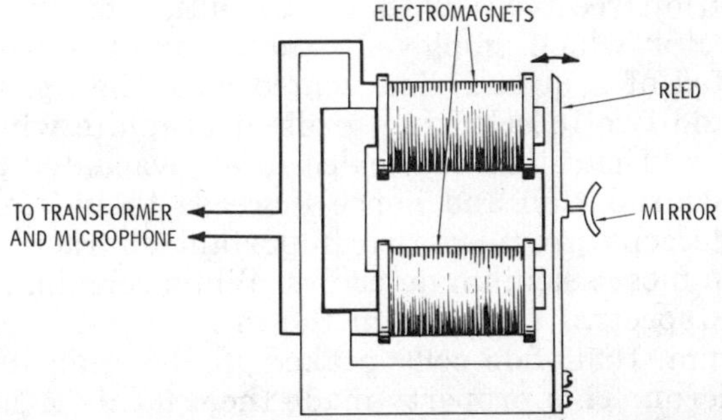

Fig. 1-6. Rankine's reed-type telephone light-beam transmitter (1916).

polished pole of an electromagnet in a circuit with a telephone transmitter. While Rankine was impressed with the prospect of locating the microphone, and hence the operator, away from the light-beam transmitter in such a system, he found that the frequency response of the system was unsuitable for reasonable voice reproduction.

Rankine employed a receiver essentially unchanged from Bell's selenium cell, battery, and telephone-receiver circuit of 1880. Although vacuum-tube triode amplifiers were available, he found the noise level of the detectors to be so high that amplification was impractical. Nevertheless, Rankine achieved good communication ranges with his equipment. For example, two carbon-arc transmitters and two 50-cm projectors of poor optical quality gave a two-way communication range of 2.4 kilometers (1.5 miles), even when ". . . considerable mist intervened." With the carbon arc replaced by light from the sun, he reported that "the faintest whisper could be heard at this distance."

Rankine's contribution to light-beam communication ended with the war, but his brief investigations were complete and were of the highest caliber, considering the available apparatus. It is interesting to note that wartime shortages forced him to borrow and improvise key portions of his light-beam communicator. His treatment of such optical-communication tradeoffs as lenses versus reflectors, detector response and noise, field testing, and other practical considerations was far superior to that of many later workers in the field.

American Systems

The selenium detector used by Bell in 1880 remained basically unchanged for 37 years until 1917. In that year, T. W. Case of Auburn, New York made the first advance in optical-communication receivers with his invention of the thallofide cell, a detector which employed oxidized thallous sulfide. The cell consisted of a quartz disc, coated with the light-sensitive material, and two interlocking grids of graphite which served as electrodes. The cells were sealed in an evacuated glass tube to reduce degradation and hence increase their life.

Thallofide cells provided two important characteristics not available in most selenium detectors. While selenium detectors had a peak spectral response in the green portion of the visible spectrum, thallofide cells peaked in the near infrared at about 1 micron. This property made them ideal for use in covert communication links employing invisible near-infrared radiation. The second important property of the cells was a high dark resistance of up to 500 megohms. This made them ideal for use with external vacuum-tube amplifiers.

Case developed a novel circuit for the thallofide cell which was used in a demonstration before United States military representatives in October 1917. The circuit, which is reproduced in Fig. 1-7, operates as a relaxation oscillator in which

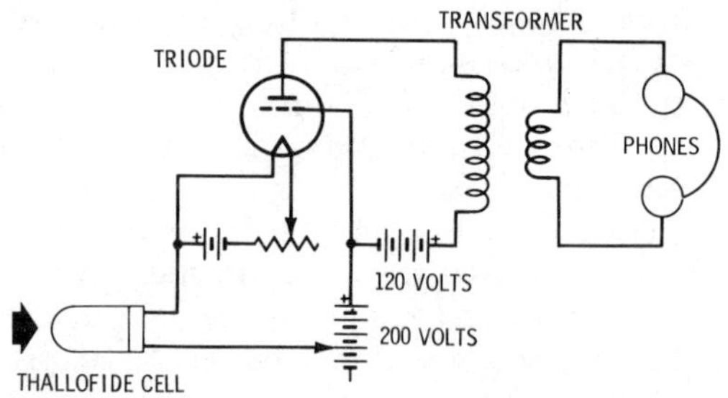

Fig. 1-7. Case's relaxation oscillator thallofide receiver circuit (1917).

a triode serves as a negative-resistance device. The circuit produced an audible tone where frequency varied with the intensity of light illuminating the thallofide detector. The cell was placed at the focus of a 61-cm (24-inch) parabolic reflector.

The 1917 demonstration took place over a range of 29 kilometers (18 miles) through smoky atmospheric conditions. The transmitter consisted of a 150-cm (60-inch) searchlight covered by an infrared filter. Code messages were successfully sent with the apparatus.

Case developed a prototype system for military use by the end of World War I. The system had a pair of 20-cm (8-inch) parabolic reflectors at each of two transceivers. An incandescent lamp served as an optical source, and an infrared filter provided covert operation. The system was battery powered, could be carried by two men, and provided a two-way code transmission range of 6.4 kilometers (4 miles). For a time, in fact, Case Research Laboratory manufactured and marketed the thallofide cell.

Case also developed a voice-modulated oxyacetylene flame. The source produced 150 candlepower and was collimated by a 61-cm (24-inch) parabolic reflector. On a clear night and with a 61-cm collector at the receiver and with a three-stage vacuum-tube amplifier, good-quality voice transmission could be achieved over a range of 8 kilometers (5 miles). Case discontinued work on the apparatus when the war ended, having concluded that its usefulness was limited to military roles.

W. W. Coblentz of the United States National Bureau of Standards also developed an experimental optical-communication system during World War I. Like Case, Coblentz contributed to the state of the art by developing a new kind of detector. He found that molybdenite provided higher sensitivity than selenium, and he successfully transmitted coded signals over a 4.8-kilometer (3-mile) path by employing a detector made from the new material. The transmitter consisted of a 250-watt tungsten lamp placed at the focal point of a 28-cm parabolic reflector. A 16-cm reflector was used at the receiver.

Like Case, Coblentz employed a three-stage vacuum-tube amplifier to enhance reception, but Case's thallofide detector enjoyed far more success than the one devised by Coblentz.

The Coblentz receiver converted the radiation from the transmitter into a series of pulses, suitable for amplification, by means of a chopping wheel placed before the molybdenite crystal detector. The resulting pulses had a width of two milliseconds. Coblentz also devised a variable-pitch receiver system which employed a potassium-hydride phototube as a detector.

Germany's *Lichttelephonie*

H. Thirring of Austria devised a light-beam communicator which was actually put into production near the end of World War I. The system consisted of a voice-modulated carbon arc installed within a 36-centimeter reflector. Inductive coupling was used to connect a carbon microphone to the arc. Other light sources, including glow-discharge lamps, were also used, but they gave less range capability. The receiver consisted of a small (1-mm^2 active area) selenium cell mounted at the focus of a 22-centimeter lens. An amplifier containing four triode tubes boosted the received signal for output to a headphone set.

Thirring's system was relatively advanced for its time. It had a voice range of 8 kilometers (5 miles), and greater ranges were available with a buzzer-produced tone which could be used to send code or provide a tranmission alert signal. The system was powered by a portable gasoline-operated generator. Beam widths were very narrow (1 degree for the transmitter and less for the receiver).

POST WORLD WAR I DEVELOPMENTS

Light-beam communications research slowed to a near standstill with the conclusion of World War I. Important developments, however, occurred in a related field—the reproduction of sound from motion-picture optical tracks. The technology of sound-accompanied motion pictures involves the constituents of a simple, short-range light-beam communication system.

By means of a modulated light source, sound is recorded on a strip of film adjacent to the actual photographs comprising the motion picture. The light source causes the optical density of the developed film to vary in exact proportion to the original sound waves. Therefore, by the passage of a steady-state light through the film, the recorded information can be converted back into sound. This is accomplished by allowing the modulated light to strike a sensitive detector. Motion-picture-development engineers of the period experimented with both mechanical and electronic modulation methods, light sources, and detectors.

Very little light-beam communications research was conducted in this period, although experimenters and amateur radio enthusiasts occasionally constructed working light-beam communicators. Some were strictly demonstration devices intended to attract customers to a store or to let visitors to an exhibition actually talk over a light beam. Others were relatively sophisticated devices intended for short- to medium-range communication usually between two fixed points.

WORLD WAR II DEVELOPMENTS

With the onset of World War II, military applications revived light-beam communications research. Both the Allied and the Axis sides produced a variety of systems, and the most important ones will be described here.

German *Lichtsprechers*

During World War II, the Axis powers, and Germany in particular, had a head start in the development of light-beam communicators. The German army extensively employed three types of communicators during the war; these were the so-called *Lichtsprechers*. Some were placed in service as early as 1935. They were manufactured by the Carl Zeiss optical firm, and all possessed outstanding design, construction, and operational features. The *Lichtsprecher* series was designated by model numbers indicating the transmitter and the receiver apertures, respectively. For example, the Li 50/60 had a 50-mm transmitter aperture and a 60-mm receiver aperture.

The Li 50/60, which employed a thallous-sulfide detector, had a range of approximately 3.2 kilometers (2 miles) and had a very narrow transmitter-beam width. The system weighed 66 kilograms (30 pounds).

Besides the Li 50/60, there were two longer-range *Lichtsprechers*, the *Lichtsprecher* 80 and the Lichtsprecher 250/130. These two *Lichtsprechers* employed 1-mm diameter lead-sulfide detectors in the receiver systems. The cells could be cooled with dry ice to enhance their sensitivity. The small size of the cells reduced the amount of unwanted background illumination falling upon their active surface. For covert operation in darkness, all three *Lichtsprechers* could be equipped with red or infrared filters. The two longer-range units will now be described in more detail.

Lichtsprecher 80—In 1942 Allied forces seized several Li 80 units in Africa. These systems consisted of two tubes, one housing the transmitter and the other the receiver, mounted on a tripod. The system weighed 119 kilograms (54 pounds). A novel mechanical modulation technique was employed, and it is shown in Fig. 1-8.

The modulation unit consisted of two prisms, an electro-magnetically activated armature, a tungsten lamp, and various optical components. In operation, audio signals were amplified and fed into the armature's coil. The armature then moved in direct proportion to the incoming audio signals. A small right-angle prism was connected to the armature so that one of its sides almost touched a mesa formed on a larger

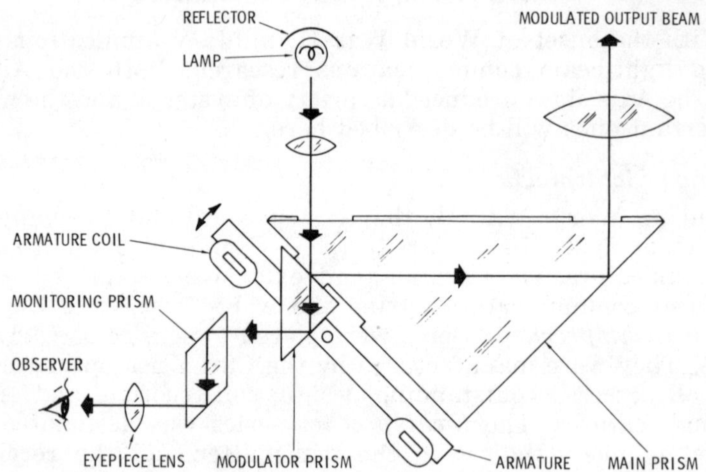

Fig. 1-8. German *Lichtsprecher* 80 modulator system (1935).

right-angle prism. As can be seen in Fig. 1-8, a light beam projected into the prism normally experienced total internal reflection at the mesa—the air interface—and was subsequently reflected out of the prism unchanged in intensity (less a small loss factor caused by absorption).

When the armature caused the small prism to contact the mesa of the larger prism, however, total internal reflection no longer occurred since the glass-air interface had been replaced by a glass-glass interface. Instead, most of the light beam directed into the large prism passed into the small prism where it was guided by a small monitoring prism into an eyepiece.

Since the amount of light passed into the small prism was dependent upon the width of the air space between both prisms, the intensity of the light beam passing out of the large prism was proportional to the movements of the armature and, therefore, to the audio signals fed into the armature's driving coil.

Due to the nonlinear modulation characteristics of this modulation scheme, voice transmissions were distorted and difficult to understand. Furthermore, the close spacing of the armature's modulating prism and the large prism made the system very sensitive to slight temperature changes. Nevertheless, this system had a very high optical efficiency since nearly 100 percent of the light collected from the source was projected when the modulator prism was separated from the large prism by only a wavelength of light.

Lichtsprecher 250/130—This communicator had both a greater range and a wider transmitter-beam width than the Li 80. United States Army tests revealed a voice-transmission range of 13.7 kilometers (8.5 miles) for two of the seized units. The half angle of the projected beam was 7 milliradians (0.4 degree). The total weight of a complete transmitter and receiver set was 309 kilograms (140 pounds).

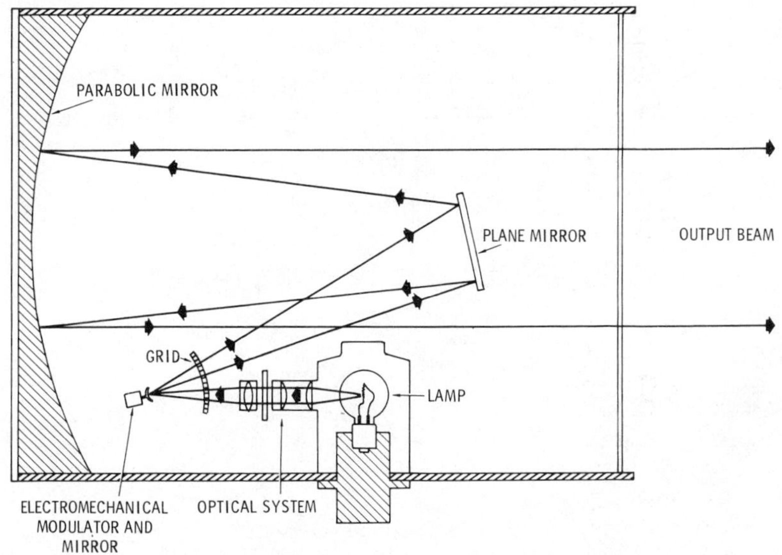

Fig. 1-9. German *Lichtsprecher* 250/130 transmitter lamp and optics (World War II).

The modulation scheme for the Li 250/130 was identical to that employed in the LI 50/60, and Fig. 1-9 is a diagram of the transmitter optics and modulator. The principle of operation is similar to that employed by Rankine, except that the opaque grids of Rankine's Photophone have been replaced by two sets of narrow precision prisms.

Fig. 1-9 shows the operation of the modulator. Voice signals are fed into a vibrating galvanometer fitted with a small mirror. Flux from a high-intensity tungsten lamp is collected by

multielement projection optics and is projected through the
first set of grid prisms where it is divided into two beams having a small deviation angle. This flux is then focused upon the
modulating mirror where it is then reflected through the second grid of prisms. From the second grid of prisms, it goes
onto a plane mirror where it is reflected toward a large 25-
centimeter concave mirror. The concave mirror forms the flux
into a narrow beam for transmission to a distant receiver
system.

Modulation occurs when angular movements of the galvanometer mirror cause the second set of prisms to recombine the
two beams produced by the first set of prisms into a single
beam of maximum intensity. Intermediate positions of the
modulator mirror cause the light to be divided between a central, combined beam and two side beams (Fig. 1-10). There-

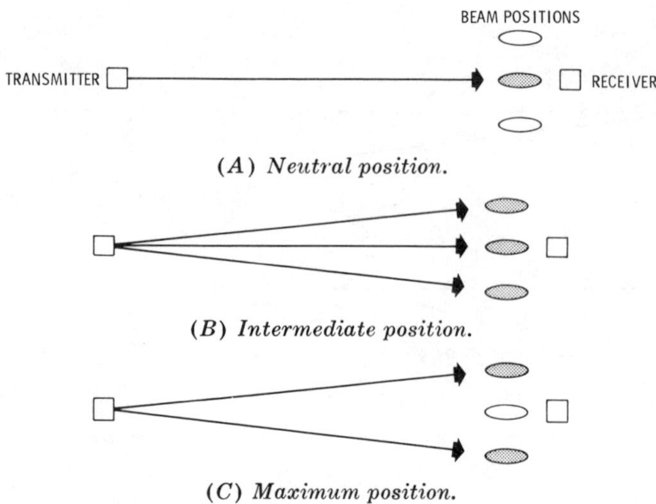

(*A*) *Neutral position.*

(*B*) *Intermediate position.*

(*C*) *Maximum position.*

**Fig. 1-10. *Lichtsprecher* 250/130 beam patterns for various modulator
mirror positions.**

fore, the magnitude of the central beam varies with respect
to the position of the modulator mirror, and high-quality voice
transmissions are made possible. An automatic gain control
feature restricts the movements of the galvanometer so that
overmodulation and the resultant distortion are prevented.

The Li 250/130 communicator could be equipped with sun-
light-collecting optics for daylight operation. U.S. Army investigators found this system to have superior optical-align-
ment and construction features.

Japanese Light-Beam Telephone

Like the German *Lichtsprechers,* the Japanese light-beam telephone employed a mechanical modulation scheme. Fig. 1-11 is a cross-sectional view of the Japanese device, showing the most significant components.

In operation, voice signals were amplified and fed into a galvanometer to which a small mirror was attached. Flux from a tungsten lamp was passed through a filter (clear, yellow, red, or infrared), passed through a grid, collimated, and projected onto the modulation mirror. The flux was then passed through a second grid and out of the instrument via a large collimation lens. Modulation occurred when the vibrating image of the first grid was passed through the fixed second grid. Only half, or less, of the flux collected by the lamp optics was contained in the transmitted beam because of the blocking action of the two grids.

The receiver system of the Japanese communicator was unique since the transmitter lens served as a receiver collection lens. As can be seen by referring to Fig. 1-11, flux from a distant transmitter was collected by the lens, reflected by the back side of the second grid, and focused onto a phototube. The signal from the tube was amplified and fed into a headphone set. Since only 50 percent of the received flux was reflected by the grid, an auxiliary receiving mirror was provided. If the received signal was weak, the operator simply placed the auxiliary mirror over the back side of the second grid and all the collected flux would be focused upon the phototube.

The Japanese communicator weighed 234 kilograms (110 pounds) and had a typical night range of about 2.4 kilometers (1.5 miles).

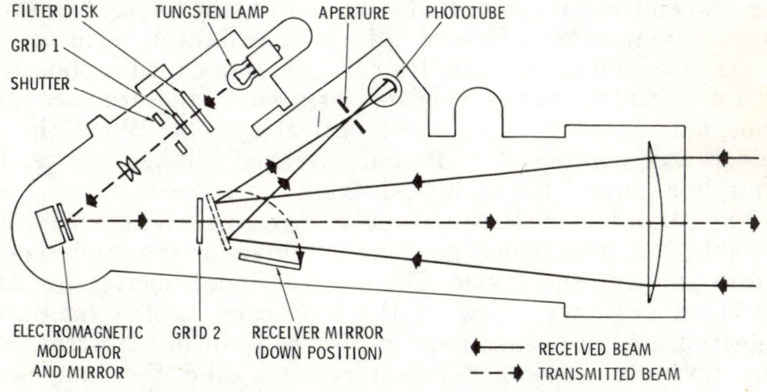

Fig. 1-11. Japanese Photophone (World War II).

Italian Photophones

The Italians developed two light-beam communication systems during World War II. Both systems were distinguished from those employed by the Germans and the Japanese in that direct modulation of a tungsten source, rather than mechanical modulation, was used. One system employed lenses and the other, concave mirrors.

The relative advantages and disadvantages of directly modulated tungsten systems are discussed in detail elsewhere in this book. Suffice it to say that directly modulated tungsten systems are generally simpler and more rugged than mechanically modulated systems but that speech modulation is more difficult and thus relatively inefficient.

American Developments

Due to its early entry into the war, Germany initially held a significant lead over the United States in the field of light-beam communications. Even though the German forces had been using the *Lichtsprecher* since 1935 and the U.S. Optical Division of the National Defense Research Committee did not begin work on proposed light-beam communicators until 1941, the United States was eventually able to devise a variety of usable light-beam communicators. Considerable work was performed in improving traditional on-off blinker-code optical-transmission devices. Both high-speed mechanical shutters and sophisticated electronic-switching circuitry were employed to produce blinkers having transmission rates of up to twelve words per minute.

Since light-blinker transmission systems could be detected by the unaided eye, infrared filters were often used to render the transmitted beam invisible to the unaided eye. The invisible beam was transformed into visible light by one of two kinds of image-conversion devices. One was called a phosphor-button receiver, and it consisted of a small infrared-sensitive phosphor screen and a simple optical system. When the receiver was pointed at a distant infrared blinker source, the phosphor glowed for each flash from the source.

The phosphor-button receiver was inexpensive, sturdy, and reliable, but it provided no image or view of the landscape to aid in pointing the device. The electron image-conversion tube provided an actual view of the landscape, which facilitated detection of the flashes from a distant light-blinker transmitter. Both the phosphor-button receiver and the image-converter tube are described in more detail in Chapter 4.

American work in voice-modulated light beams resulted in the development of the cesium-vapor lamp near the end of the war in 1944. This lamp emitted at 852.1 and 894.4 nanometers in the near infrared. The small amount of visible radiation emitted by the lamp was easily blocked by an infrared filter. These lamps converted approximately 21 percent of the incoming electrical power into usable near infrared. Since incandescent lamps required considerably more filtering to achieve the same degree of visual security as a cesium lamp, the new lamps were nearly seven times as efficient as conventional incandescent lamps.

The first demonstration of two-way voice and code transmissions with cesium lamps took place October 18, 1944 between a Navy ship on the Atlantic Ocean and a shore installation. Other types of cesium lamps were employed in a variety of applications, including a 500-watt unit built for the Air Force and designed to be used without collimating optics in order to give a 360-degree horizontal coverage.

POST WORLD WAR II DEVELOPMENTS

Bell Telephone Laboratories considered various aspects of light-beam telephone links after the war. For example, R. V. L. Hartley, a Bell Laboratory scientist, authored an unpublished laboratory memorandum entitled "A Theoretical Survey of Communication by Guided Light" in which he considered the possibility of employing transparent rods, internally reflecting pipes, and lenses to guide a modulated light beam. Other Bell Laboratories scientists described optical repeater systems in similar unpublished papers.

The invention of the transistor by three Bell Laboratories scientists in 1949 sparked a major research program in semiconductor electronics. In 1951 Howard Briggs, James Haynes, and William Shockley, all three employed by Bell Laboratories, discovered that pn junctions fabricated in silicon and germanium emitted infrared radiation when forward biased. The scientists, who were granted U.S. Patent 2,683,794 for their invention, performed a series of tests which revealed a narrow spectral bandwidth and a fast radiation rise time from the forward-biased infrared emitters.

In 1954 Yves A. Rocard of Paris, France described several unique optical-communication systems based upon infrared-emitting diodes made of silicon, germanium, and other semiconductor infrared sources. In his U.S. Patent 3,111,587, Rocard described new methods of constructing efficient semiconductor sources, a technique for two-way communications which utilizes only one light source, and a variety of infrared voice communicators.

In 1962 J. I. Pankove of RCA Laboratories reported the development of infrared-emitting semiconductor diodes having much higher efficiency than those made from silicon and germanium. Since diodes made from the latter materials were very inefficient, they were not viable candidates for a practical light-beam communication system. Pankove's diodes were made from gallium arsenide (GaAs), a semiconductor compound having superior optoelectronic properties..

In 1962 scientists at Lincoln Laboratory, Massachusetts Institute of Technology, used a GaAs diode to send a 4-MHz television signal to a receiver 84 meters (275 feet) away. Early in 1963, the same team of scientists sent both audio and television signals from the crest of Mt. Wachusett, Princeton, Massachusetts to the roof of Lincoln Laboratory, a distance of 55.6 kilometers (30 nautical miles). The infrared source was a GaAs diode cooled with liquid nitrogen and emitting at 840 nanometers. The transmitter optics produced a 2-milliradian beam. The receiver was a photomultiplier tube mounted at the focal point of a searchlight reflector having a 1.5-meter (5-feet) diameter.

The MIT experiments with infrared-emitting diodes remain a major milestone in light-beam communication technology. Soon, other research groups demonstrated similar GaAs-diode communication systems, and Texas Instruments introduced a GaAs infrared emitter in 1962. The firm also developed several types of portable GaAs communicators for military and demonstration purposes.

THE LASER

The invention of the laser in 1960 opened up a new range of possibilities for light-beam communications. Lasers are discussed in detail in Chapter 3, and suffice it to say at this point that the coherent beam emitted by the appropriate kind of laser has the potential of transmitting a half-million television signals *simultaneously!*

Laser technology has advanced much faster than the technology required to modulate and demodulate such an enormous amount of information, and until several years ago many scientists felt that practicable laser communications would be limited to military roles and possibly to very-short-range civilian atmospheric links. Because of the sophistication of microwave-transmission techniques, little hope was held for space applications.

A chief factor in the pessimism was the unpredictable nature of the transmission medium, the atmosphere. In May 1963, a group of engineers from Electro-Optics Systems transmitted a voice-modulated helium-neon laser beam (632.8 nanometers) from Panamint Ridge near Death Valley to the San Gabriel Mountains near Pasadena, California, a range of 190 km (118 miles). This experiment remains the distance record for voice-modulated light-beam communications, though advancing technology now permits even greater range capability. Despite the success of the 1963 demonstration, such common atmospheric constituents as turbulence, rain, snow, fog, dust, clouds, and pollutants seriously impair the prospects of practical long-range atmospheric links.

Courtesy Bell Laboratories

Fig. 1-12. Light emerging from the ends of glass fibers.

Fortunately, the recent development of very-low-loss optical transmission fibers (Fig. 1-12) and of new kinds of efficient, long-lived semiconductor lasers (Fig. 1-13) and detectors has given renewed impetus to practicable light-beam communications. AT&T long ago ruled out atmospheric links for practicable intracity and intercity telephone links. Now, with the development of low-loss optical fibers at Corning, Bell Laboratories, and elsewhere, AT&T is funding a major optical-fiber communication research effort.

Until recently, the best available fibers had losses of about 1 dB/meter—far too high for more than very-short-range links. Fibers have now been made with a loss of less than 2 dB/km. Even a fiber with a loss of 5 dB/km will attenuate only 50 percent of an optical signal in 0.6 km (2000 feet)—

Fig. 1-13. High-efficiency LOC injection laser diode.

significantly less than conventional wire-system losses. This means repeaters can be spaced farther apart in an optical system. For example, one proposed fiber-optic system using 5-dB fiber would require repeaters only every 13 km, while conventional wire systems require 8-km intervals.

The use of low-loss fibers means that, with existing technology, a 6-mm cable of hair-thin glass fibers can carry the same volume of communications as thousands of conventional telephone cables. The glass cables are much smaller than metal ones and could be run in the spaces in existing conduits.

The major drawbacks of optical fibers include inefficient manufacturing techniques, inability to transmit electrical power through the fiber, and inadequate installation and repair techniques.

CHAPTER 2

Principles of Light-Beam Communications

A light-beam communication system requires an optical source, a means for modulating the source, and an optical receiver. Frequently, optical components are included at both the transmitter and the receiver in order to increase the quantity of radiation transferred from the transmitter to the receiver. This transfer efficiency directly affects the range of the system. The transmission medium also affects the range of the system as do such factors as the transmitter power and the receiver sensitivity.

This chapter discusses many of the factors that affect the operation of a light-beam communication system. By definition, light refers to those optical wavelengths that are visible. But for the purpose of this book, the term "light-beam communications" refers to all systems that utilize an optical carrier. The optical spectrum, as can be seen in Fig. 2-1, includes those wavelengths ranging from the ultraviolet to the far infrared.

A better appreciation of which wavelengths are best suited for optical communications can be had by considering some of the sources of optical radiation employed in optical communicators. There are an impressive number of factors that must be considered in choosing an optical source (e.g., detector availability, atmospheric transmission, physical configuration of the source, etc.), and some of them will be discussed here.

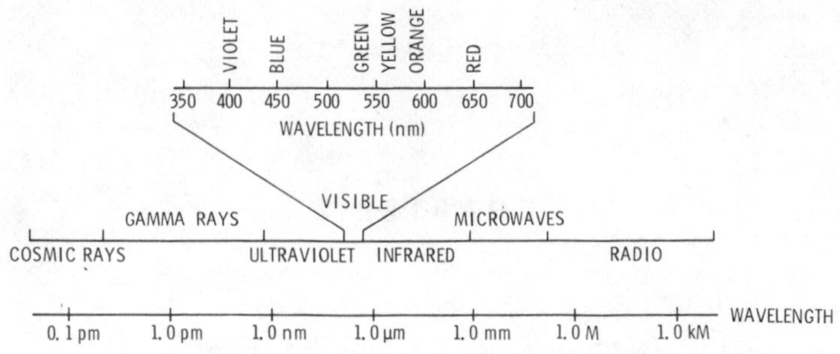

Fig. 2-1. Electromagnetic spectrum.

LIGHT SOURCES

Numerous radiation sources have been used in optical-communication links. These include the sun, arc lamps, gaseous discharge tubes, incandescent filaments, semiconductor emitters, and various kinds of lasers. As noted in detail in the preceding chapter, the sun is a practical and convenient source. Arcs are efficient light emitters, and the carbon arc is among the brightest of light sources. Efficient modulation of arcs, however, can be difficult. Gaseous discharge tubes are easily modulated, but they radiate considerably less light than arcs and other sources. Incandescent lamps are economical and readily available, but the mass of their filament means they possess relatively slow rise and fall times. Therefore, 100-percent modulation is impossible at voice frequencies and a considerably smaller modulation rate must be employed.

Semiconductor emitters include the light emitting diodes (LEDs). These are true solid-state light sources with an impressive modulation capability. Also, LEDs are narrow-wavelength emitters and are, therefore, well suited for operation in optically noisy environments.

Laser sources come closest to approaching the requirements of an ideal radiation source for optical communications. The radiation from many lasers is partially or even almost completely coherent. This, coupled with the relatively narrow beam of most lasers, means that extremely narrow beam widths are possible.

This survey of light sources is necessarily brief, and the subject is covered in detail in Chapter 3. It is important to know that a variety of sources has been employed in optical communications and that each of the basic source categories, particularly lasers, is represented by a number of variations and modulation methods.

LIGHT DETECTORS

The first light-beam communicator used a selenium detector, and selenium remained popular until better detector substances were discovered during and after World War I. Although selenium is not used in the detectors of commercial light-beam communicators, the material is sometimes used by hobbyists and experimenters to make experimental communication links.

Selenium is characterized by two electrical properties. When the element is exposed to light, the resistance falls dramatically. Light also induces an electromotive force in selenium. These two phenomena are called, respectively, the photoresistive and the photovoltaic effects. These two classes of detectors, as well as several others, are considered at length in Chapter 4. For now, it is only necessary to realize that for most sources of optical radiation there exists an appropriate detector. This is a fortuitous occurrence, for detectors tend to have a peaked spectral sensitivity and emitters tend to have a peaked spectral output.

Typical photoresistive detectors include cadmium sulfide (CdS) and cadmium selenide (CdSe) cells. While these detectors are economical and have excellent sensitivity to the light of invisible and near-infrared wavelengths, respectively, they have very slow response times and exhibit fatigue effects.

Photovoltaic detectors are characterized by faster response times than photoresistive devices. In some cases they are usable over very wide ranges of incident light density, thus making them useful for operation in the presence of relatively bright background light. Silicon is the most common photovoltaic material.

An important type of detector is the photoconductor. Like some photovoltaic units, photoconductive detectors can be operated in the presence of normally undesirable background light. In fact, a photovoltaic detector can be operated in a photoconductive mode by simply applying a *reverse*-bias voltage. Because of their very fast response time, photoconductive detectors are important in semiconductor LED and laser communication systems.

Another important detector is the avalanche photodetector, a specially fabricated pn device that combines high sensitivity with an exceptionally fast response time. No other existing semiconductor detector has both of these characteristics simultaneously, and only certain photomultiplier tubes have greater sensitivity.

Because of their very high sensitivity, photomultiplier tubes are used in very-long-range optical-communication links. While their sensitivity is high, they are very fragile, require a high bias voltage for proper operation, and are difficult to operate in the presence of ambient light.

An important detector for systems employing near-infrared radiation is the *image converter,* a generic term for a family of devices which convert invisible infrared into visible light. Some image converters are as simple as a screen of phosphor which glows orange or green in the presence of infrared. Others employ evacuated viewing tubes and a high-voltage power supply. The image converters and the other detectors briefly described here are discussed in Chapter 4.

MODULATION OF LIGHT BEAMS

There are numerous techniques that can be used to convey information from one point to another with a light beam. The simplest is to periodically interrupt the beam, using a prearranged code. More-sophisticated modulation methods vary the frequency, amplitude, polarization, intensity, or phase of a light beam, using either analog or pulse-modulation formats.

A common modulation technique is to vary the electrical current passing through the radiant source in step with an incoming signal. This method works well with incandescent lamps, some arc lamps, light emitting diodes, and some lasers.

While direct modulation of the radiant source is generally uncomplicated, some sources cannot be directly modulated. The flux from these sources must be modulated indirectly by passage through any of a variety of external modulators.

One type of external modulator is the Kerr cell. Shown in Fig. 2-2, a Kerr-cell modulator consists of a liquid whose po-

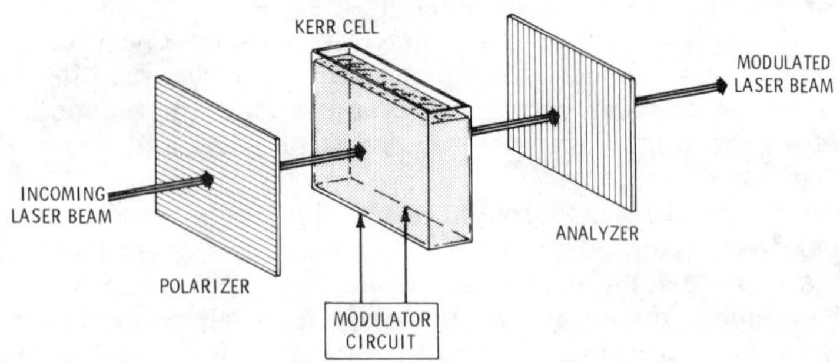

Fig. 2-2. Construction of a Kerr-cell modulator.

larization plane can be altered by an electrical field and two crossed polarizers. Normally no light is emitted by the modulator; but, when the Kerr cell is activated, its polarization plane rotates up to 90 degrees and permits light to pass through the system.

A more common external-modulation method utilizes the Pockels effect. Certain crystals, such as ammonium dihydrogen phosphate (ADP) and potassium dihydrogen phosphate (KDP), vary their polarization plane when an electrical field is applied. Fig. 2-3 shows a modulator utilizing the Pockels effect. Operation of the modulator is similar to the Kerr-cell modulator described. Some Pockels-effect modulators employ two crystals.

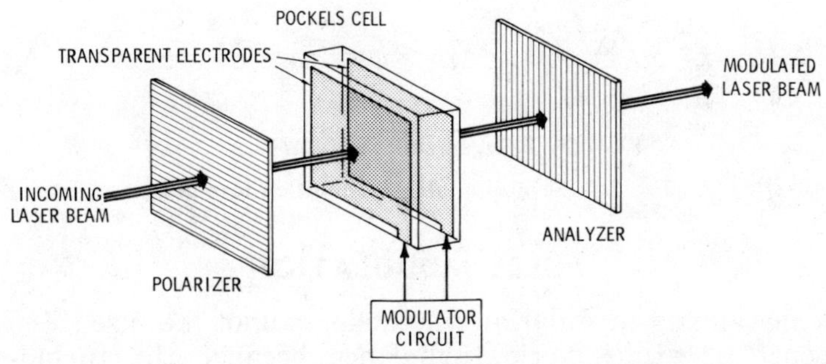

Fig. 2-3. Construction of a Pockels-cell modulator.

Mechanical modulators can also be used to modulate a light beam. Several such systems are described in Chapter 1. Other modulation techniques are described in Chapter 3.

ANALOG MODULATION

The various modulation methods described above can be used to encode information onto a light beam in either an analog or a pulse format. The simplest analog-modulation format causes the amplitude or intensity of the light beam to vary in direct proportion to an incoming analog signal, such as voice. This format is called amplitude modulation (a-m) or intensity modulation (im).

Analog frequency modulation (fm) modulates the optical frequency (wavelength) of the source. Analog phase modulation (pm) modulates the phase of the radiation emitted by the source. And analog polarization modulation modulates either the linear or the circular polarization of the beam from the source. Fig. 2-4 reviews several analog-modulation formats.

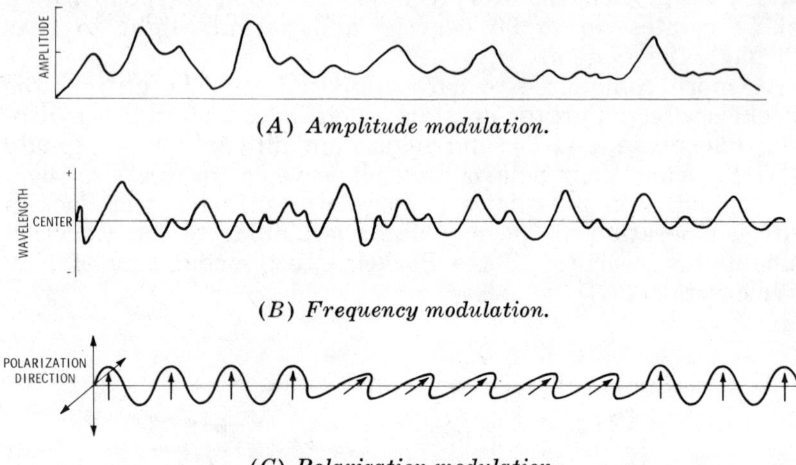

(A) Amplitude modulation.

(B) Frequency modulation.

(C) Polarization modulation.

Fig. 2-4. Several analog-modulation formats.

PULSE MODULATION

Some analog-modulation methods cannot be used in the presence of atmospheric turbulence because the turbulence will add considerable noise to the transmitted signal. Also, some sources cannot be readily modulated with analog methods. Pulse modulation eliminates both these problems nicely. Various types of pulse modulation can be used to transmit an optical signal through a turbulent medium with little or no adverse effect. And, if a threshold-discrimination circuit is incorporated into the receiver, the received signal will have a constant output until the received signal level falls below the threshold level. Semiconductor infrared emitters and diode lasers are particularly well suited to pulse modulation.

Numerous kinds of pulse modulation have been devised, some of which encode the transmitted signal so that reception by unauthorized parties is difficult or even impossible. One of the simplest pulse-modulation methods is pulse-frequency modulation (pfm). In this method, a radiant source is modulated with a fixed pulse-repetition rate, called the carrier frequency. An incoming analog or digital signal causes the carrier frequency to vary within a predetermined range. The receiver circuitry filters out the carrier frequency and processes only the modulated signal.

Pulse-amplitude modulation (pam) sets the amplitude of a train of regularly spaced pulses according to the amplitude of an incoming signal. Pulse-width modulation (pwm) varies the width of regularly spaced pulses according to the amplitude of an incoming signal.

Pulse-rate modulation (prm) resembles pfm in that the number of regularly occurring pulses per unit time interval is varied according to the amplitude of an incoming signal.

Pulse-modulation techniques are ideal for transmitting digital data, and an interesting variety of digital-modulation methods has been devised. For example, the carrier amplitude can be set at some maximum value to represent one bit of binary data and at some minimum value to represent a zero bit of binary data. Techniques to implement this form of pulsed-digital modulation are called pulse-coded modulation (pcm).

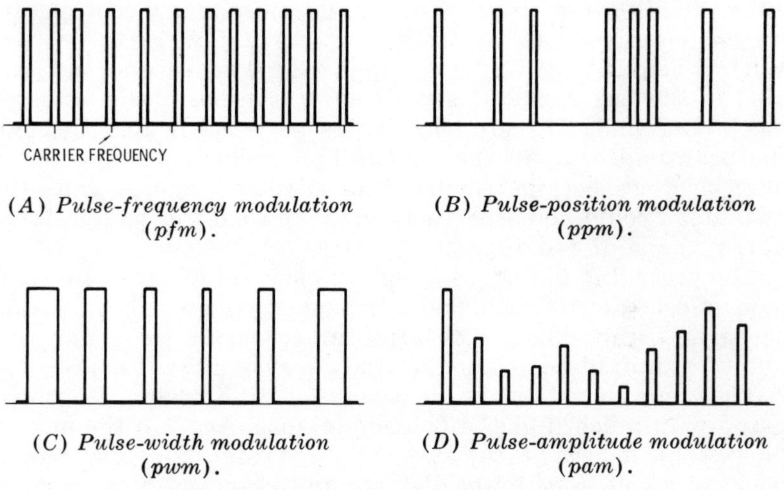

(A) *Pulse-frequency modulation (pfm).*

(B) *Pulse-position modulation (ppm).*

(C) *Pulse-width modulation (pwm).*

(D) *Pulse-amplitude modulation (pam).*

(Fig. 2-5. Several pulse-modulation formats.

Pcm can also be used to send binary data by assigning a zero bit to one optical frequency or pulse-rate frequency and a one bit to another frequency or pulse rate. These techniques are called pcm/fm or pcm/fsk.

Fig. 2-5 reviews several common modulation formats. Numerous other modulation schemes for using pulses to encode analog or digital information have been devised. Those described here are merely representative of the techniques available.

ATMOSPHERIC PROPAGATION

Several years ago I participated in a field test of a sophisticated laser illumination and detection system conducted by the Laser Division of the Air Force Weapons Laboratory. One objective of the test was to photograph the beam pattern from a 50-mW helium-neon laser as it illuminated a large target board. The board was painted white and sprayed with tiny glass beads to enhance its reflective properties.

After having passed through a kilometer of relatively still air on a cold evening, the cross section of the laser beam was a spectacular sight. The beam had spread to nearly the diameter of the target and appeared as a bright red, circular, net-like structure which constantly varied in appearance. The overall sensation was very similar to observing the reflection of the full moon in a slightly disturbed surface of water.

Other phenomena could also be observed. For example, the entire scintillating circle of red would move a few centimeters in various directions at frequencies of a few hertz or less. When a vehicle drove down an unpaved gravel-access road parallel to the test range, dust drifted by the beam and holes of black surrounded by circular fringe patterns floated through the beam pattern. As the vehicle approached the target, the dust shadows became smaller and smaller. All the time, the dust cloud could be observed to move toward the screen like a narrow cone of red fog.

The preceding narrative illustrates several of the more common effects of the turbulent atmosphere upon a light beam. These and other effects can significantly impair the range capability of a light-beam communication system.

The beam of a light-beam communication system is propagated with much higher efficiency in space than in the atmosphere. This is because the atmosphere is essentially a turbulent medium which constantly distorts and rearranges the energy distribution within the beam. Worse, particles of matter, which may range from various kinds of precipitation to microscopic specks of particulate matter, are dispersed throughout the atmosphere and may act to further disrupt the passage of a light beam by blocking it. Finally, the atmosphere itself acts to absorb, and therefore attenuate, a light beam no matter what the visibility range happens to be.

Since atmospheric transmittance is a vital aspect of a light-beam communication system designed to operate in the open, it is important that any book on the subject discuss this topic. We shall begin by discussing the atmosphere itself, and then we shall describe several of the most important phenomena which affect the passage of a light beam through the open air.

THE ATMOSPHERE

The atmosphere is a mixture of several gases; nitrogen and oxygen are the predominant constituents, having percentages of 78.088 and 20.949, respectively. The next most common gases in the order of their prevalence are argon (0.93%) and carbon dioxide (0.033%). Traces of neon, helium, methane, krypton, nitrous oxide, carbon monoxide, xenon, hydrogen, and ozone are also present but rarely in concentrations of more than a few parts per thousand or even million.

ATMOSPHERIC ABSORPTION

The composition of the natural atmosphere is frequently varied by the presence of water vapor and a great variety of natural and man-made pollutants. Water vapor is of great importance to light-beam communications since it selectively absorbs certain optical wavelengths far more than others. For example, the 940-mm wavelength emitted by high-efficiency gallium-arsenide light emitting diodes compensated with silicon is far more susceptible to attenuation by water vapor than the 905-mm wavelength of conventional nonsilicon-compensated gallium-arsenide LEDs.

Certain naturally occurring atmospheric gases also attenuate light waves. For example, carbon dioxide and nitrous oxide absorb certain narrow bands of optical radiation throughout the infrared portion of the spectrum beginning at about 1.4 microns.

Regions that suffer little or no atmospheric attenuation due to selective spectral absorption are termed *windows*. Fig. 2-6,

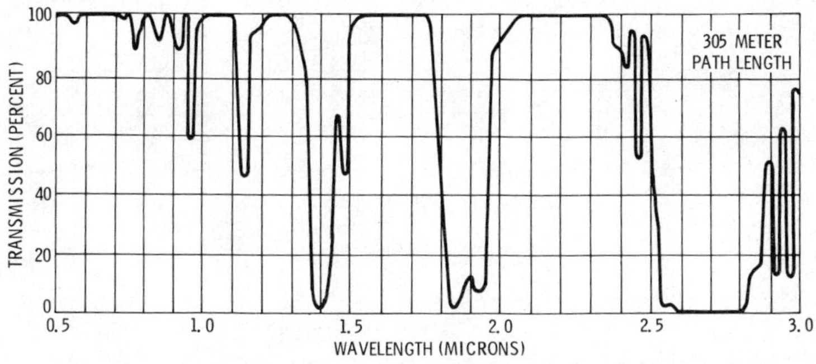

Fig. 2-6. Atmospheric spectral transmittance from 0.5 to 3 microns.

a graph of atmospheric transmittance from 0.5 to 3 microns, reveals many such windows (e.g., 0.6-0.7 micron, 0.82 micron, etc.).

Even atmospheric windows are subject to major attenuation. Smoke, haze, fog, dust, automobile particulate emissions, and other particles are not nearly so selective as water vapor, carbon dioxide, and nitrous oxide. Instead, they tend to absorb, reflect, or scatter optical radiation of all wavelengths.

There are two primary types of atmospheric scattering: Rayleigh and Mie. Rayleigh scattering pertains to molecular-sized particles and is of less concern to light-beam communicators than is Mie scattering. Rayleigh scattering is responsible for the deep blue color of a clear sky. This is because the short, blue wavelengths of solar illumination are scattered far more than the longer wavelengths.

Mie scattering pertains to particles that are large in comparison to the wavelength of a passing beam of light. Mie scattering is directly responsible for the visibility ranges provided by the United States Weather Service for the benefit of aviators and others requiring such information (such as light-beam communicators).

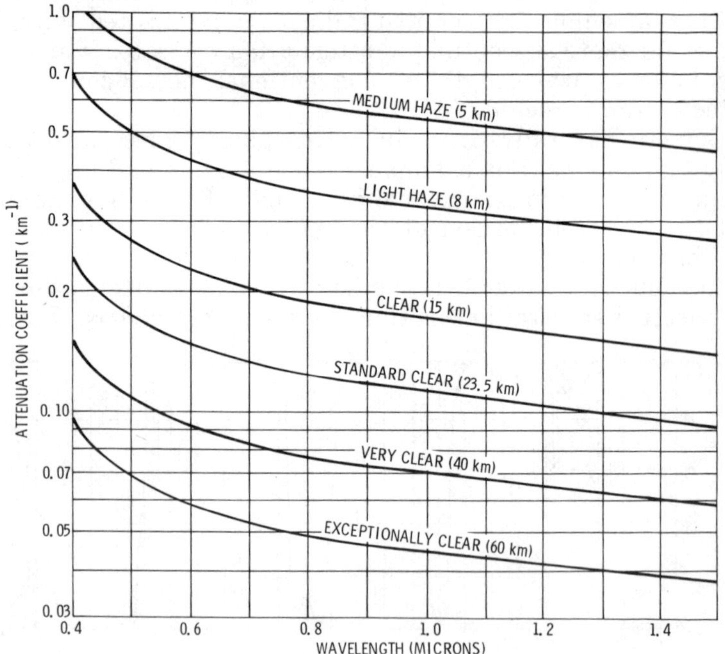

Fig. 2-7. Attenuation coefficent versus wavelength (sea level).

Meteorologists have defined atmospheric visibility (meteorological range), hence transmission range, as that horizontal distance for which the contrast transmission is 2 percent. They have also assigned the following terms to specify atmospheric visibility at sea level:

Condition	Visibility (km)
Haze	3
Medium Haze	5
Light Haze	8
Clear	15
Standard Clear	23.5
Very Clear	40
Exceptionally Clear	60

These atmospheric conditions and resultant visibility ranges have been converted to give attenuation coefficients for a range of wavelengths. Fig. 2-7 is one such graph showing the attenuation coefficient as a function of wavelength.

ATMOSPHERIC TURBULENCE

The atmosphere is a dynamic, ever-changing medium. Even on the stillest of days, subtle mixing and movement of the atmospheric contents occur. The dynamic nature of the atmosphere and the variances in its refractive index give rise to a variety of phenomena which adversely affect light-beam communications. These include such effects as scintillation, image dancing, beam spreading, and beam steering. The atmosphere can even cause the spatial coherence of a laser beam to be badly degraded.

ATMOSPHERIC MEASUREMENTS

Numerous experiments have been conducted to measure the effect of the atmosphere on a light beam. One of the most interesting experiments was conducted in 1964 by the National Bureau of Standards. The experiment took place over two separate path lengths, one of 15 km (9.3 miles) and the other of 145 km (90 miles). Similar atmospheric conditions prevailed when tests were conducted over both paths.

A helium-neon laser having an output of 5 milliwatts at 632.8 nanometers was employed as an optical source. The laser beam was collimated with an 11.4-cm (4.5-inch) telescope. The beam was detected by a 7.6-cm (3-inch) lens and focused upon a photomultiplier tube.

The highly collimated beam from the laser spread to a width of 51 cm (20 inches) at 15 km, and 914 cm (30 feet) at 145 km. The beam power density at the receiver varied considerably in amplitude over both paths. The frequency of the fluctuations varied from a fraction of a hertz to several hundred hertz. Additionally, in both tests the beam wandered over an area several times its diameter.

The similarity in fluctuations over the two path lengths may have been caused by the higher average elevation of the 145-km path. Other tests have shown better transmission properties of the atmosphere at higher elevations.

OPTICAL CONSIDERATIONS

Some very-short-range optical-communication systems require no optics, but most practical communicators do require various optical elements. Chapter 5 describes many kinds of optical components applicable to light-beam communication systems.

The lens is the most common optical element in an atmospheric optical link. Lenses are used to collect radiation from a source and form it into a collimated beam. Lenses are also used to collect radiation from a distant transmitter and focus it upon a detector.

A knowledge of the size of a light source and the focal length of a lens permits the beam divergence of a source-lens combination to be readily calculated by applying the following formula:

$$\theta = \frac{d}{f} \qquad \text{(Eq. 2-1)}$$

where,

θ is the divergence in radians,
d is the width of the source,
f is the focal length.

For example, a light emitting diode having a chip 1 mm in diameter and coupled to a lens having a focal length of 10 mm will give a beam divergence of 0.1 radian. The result can be converted to degrees by multiplying by 57.3.

Equation 2-1 also applies when you are calculating the beam divergence, more properly the field of view, of a lens coupled with a detector. A large receiver beam width tends to reduce pointing errors, and a small beam width tends to reduce background irradiance on the detector.

A knowledge of the range over which a light beam is projected and its divergence permits the size of the illuminated spot to be calculated as follows:

$$d = R\theta \qquad \text{(Eq. 2-2)}$$

where,

d is the illuminated spot diameter,
R is the range to the spot,
θ is the beam divergence in radians.

Many optical transmitters impose special restrictions upon a collimating lens. This is because longer focal lengths are usually desired in order to produce narrow beam widths, but long-focal-length lenses collect less radiation from the source. The figure of merit for the collection efficiency of a lens is its f/number, sometimes referred to as the "speed" of a lens. The f/number is given by

$$f/\text{number} = \frac{f}{d} \qquad \text{(Eq. 2-3)}$$

where,

f is the focal length,
d is the lens diameter.

Fast lenses having a small f/number are often used in light-beam communicators when the source emits into a relatively wide angle.

Parabolic reflectors are also frequently used in optical communicators. A more detailed discussion of the application of lenses and reflectors as antennas in an optical-communication system follows.

OPTICAL ANTENNAS

Typical optical antennas are lenses and parabolic reflectors. The term "antenna" usually applies to the radiating or collecting device of a radio system, but in recent years it has been applied to optical communications as well.

Antenna gain is a figure of merit which describes the ability of a transmitter or a receiver to project or collect radiation in an electromagnetic communication system. Although light-beam communication systems can be specified in a variety of ways, specifying the antenna gain can be particularly useful when systems that employ similar sources and detectors are compared.

TRANSMITTER ANTENNA GAIN

The transmitter antenna gain of an optical communication is the ratio of the area illuminated by the device's radiant source without an optical antenna, to the area illuminated by the source in conjunction with an optical antenna.

To understand how transmitter antenna gain is calculated and applied, consider the case of an *isotropic* optical source. An isotropic source radiates uniformly in all directions, and an incandescent lamp approximates an isotropic radiator.

Fig. 2-8 shows that an isotropic source radiates into a sphere having a surface area of $4\pi r^2$. In Fig. 2-9 the isotropic source

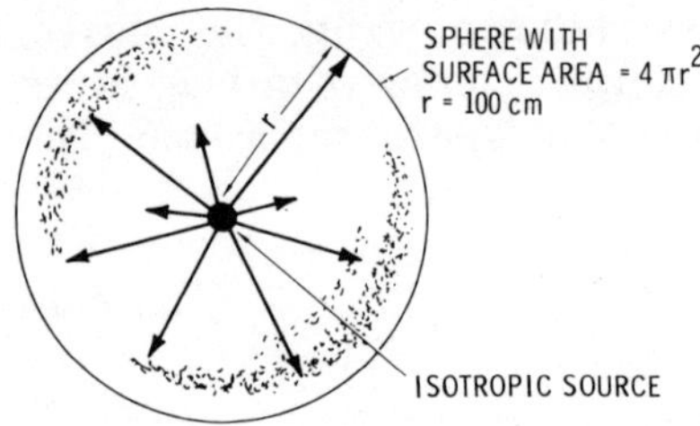

Fig. 2-8. Spherical radiation pattern of an isotropic source.

has been mounted at the focus of a parabolic reflector, and the radiant flux that formerly was spread over $4\pi r^2$ is now confined to a relatively small spot. In this case, the area of the spot is approximately πr^2 (where r is the radius of the spot) since the spot is circular and relatively small. Actually, the area of the spot is slightly greater than πr^2 since the spot is a section of a sphere, and an area calculation of a relatively large spot would have to take this into consideration.

For example, a reflector that illuminated half the area of the sphere gives a spot having a surface area of $2\pi r^2$.

We can calculate the gain of the antenna in Fig. 2-9 by making a simple ratio of the resultant spot size and the initial $4\pi r^2$ radiation pattern. Since the diameter of the resultant

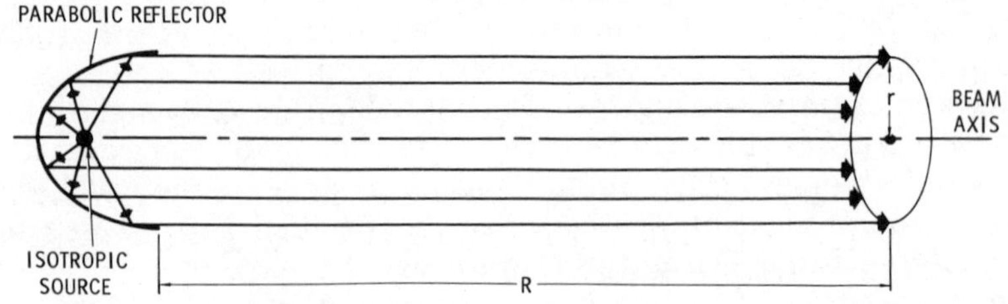

Fig. 2-9. Beam forming with a parabolic reflector.

spot is 6 cm, its area is 9π square centimeters. Since the radius of the sphere formed by the isotropic radiator in Fig. 2-8 is 100 cm, the surface area of the sphere is 2500 square centimeters. Since the antenna gain is the ratio of the latter to the former, the gain in this example is 278.

By way of comparison, gains of radio-frequency antennas are typically much lower. For example, a television-receiver antenna may have a gain of only 20. A parabolic radar reflector (a special category of microwave-frequency antenna) may have a gain of a thousand or more, but only because the short wavelength employed can be collected and focused by a parabolic dish. Most radio-frequency antennas have much lower gains than their optical counterparts. The example just given might apply to a conventional flashlight having a relatively unsophisticated parabolic-reflector antenna. Only slightly more sophisticated optical antennas can have considerably higher gains.

An example of a very high gain antenna system is a miniature incandescent lamp whose filament is placed precisely at the focal point of a good-quality parabolic reflector. The elements of this system are identical to those in the previous example, but let us assume that the lamp filament is small in relation to the size of the reflector, that the filament is exactly at the reflector focus, and that the reflector is of high optical quality.

In an actual experiment, a lamp-reflector combination such as this, illuminated a spot only 4.2 cm in diameter at a distance, r, of 10 meters. When the sphere was illuminated isotropically by the lamp alone, the surface area at 10 meters was 400π square meters. When a reflector was added, the area of the spot was 4.41π square centimeters. The ratio of the former to the latter gives a resultant antenna gain of an impressive 907,029.

The antenna gain of a radiant source that beams its radiation into a nonisotropic radiation pattern may be necessarily small. For example, a semiconductor injection laser may emit its energy into an approximately conical beam having a spread of perhaps 20 or 25 degrees. A simple f/1 lens can be used to collimate *all* this radiation into a beam having a divergence of a fraction of a degree.

It is important to note that transmitter antenna gain is the ratio of the area illuminated by the source alone to the area of the spot illuminated by the source supplemented by an antenna. Therefore, sources such as lasers, light emitting diodes containing integral reflectors, and other nonisotropic sources may give small antenna gains. This means antenna-gain figures for similar optical sources can be reasonably compared to one another, while gains for different sources (e.g., incandescent lamps and lasers) cannot be fairly compared.

Two examples will illustrate how drastically different antenna gains provide essentially similar results. In the first example, assume the source is a gallium-arsenide injection laser having a raw beam divergence of approximately 500 milliradians. Since the diameter of a spot illuminated by a source is equal to the source divergence in radians times the distance to the source, the raw laser beam will illuminate a 5-meter diameter circle at a distance of 10 meters (see Equation 2-2).

A simple f/1 lens can be used to narrow the beam from the laser into a cone having a divergence of only a fraction of a degree. In this example, assume the lens collimates the *entire* beam into a cone having a spread of 2 milliradians. From Equation 2-2, this gives a spot having a diameter of 2 centimeters at a range of 10 meters. Knowing the size of the raw and the collimated beam spots, we can calculate their respective areas and divide the former by the latter to find the antenna gain of the f/1 lens. In this case, the gain is a very high 62,500.

For the second example, assume the source is a helium-neon laser emitting a highly collimated beam having a divergence of only one milliradian. A single lens is used to collimate the beam even further, and the resultant divergence is 0.5 milliradians. The ratio of the two beam-spot sizes at a range of 10 meters gives an antenna gain of only 4, yet the beam is more tightly collimated than that of the previous example.

The examples given above convincingly illustrate the need for care when judging antenna gain of transmitter systems employing different radiation sources. Another important point in assessing antenna gain is what can be called *antenna efficiency*, and this is discussed in the next section.

TRANSMITTER ANTENNA EFFICIENCY

It is important to amend the results of an antenna-gain calculation with a similar calculation of *antenna efficiency*. In the examples given previously, it was sometimes noted that the *entire* beam was intercepted and collimated by a given optical antenna. There are many cases, however, when less than the total radiated energy from a source is captured by the transmitter antenna, and it is important to amend an antenna-gain calculation to account for such cases.

Consider, for example, an infrared-emitting diode that radiates into a complex pattern consisting of a central lobe surrounded by an outer halo. Assume that the power radiated into the halo is discounted at the expense of overall system efficiency and that only the central lobe, which has a full divergence of 0.4 radian, is employed in the system. At ten meters, this beam illuminates an area of $40,000\pi$ square centimeters. The addition of an f/1 lens reduces the beam divergence to 0.1 radian; this gives an illuminated spot area of 2500π square centimeters for an antenna gain of 16.

The *actual* antenna gain, however, must consider the ratio of transmitted to intercepted power, and in this case the f/1 lens captures less than the total central lobe radiated by the diode. Since the lobe power distribution is structured, we cannot geometrically calculate the percentage of power intercepted by the lens. Instead, an actual measurement of intercepted to total lobe power is used, and this gives the *antenna efficiency* of 64 percent. Therefore, the actual antenna gain is only 64 percent of the calculated gain (16), or 10.24.

It is important to note that the radiation from highly directional sources, such as most kinds of lasers, can be collected with a high degree of efficiency. In many cases, *all* the radiation emitted by the source can be collected. Optical antennas used to collimate the radiation from such sources will often have a collection efficiency of 100 percent.

We can now define overall transmitter antenna gain as the ratio of the spot area illuminated by the raw source to the spot area illuminated by the collimated beam, times the collection efficiency of the antenna. Thus,

$$G_{TA} = E_{TA} \left(\frac{S_u}{S_c}\right) \qquad \text{(Eq. 2-4)}$$

where,

G_{TA} is gain of the transmitter antenna,
E_{TA} is efficiency of the transmitter antenna,
S_u is spot size of the uncollimated beam,
S_c is spot size of the collimated beam.

Equation 2-4 can be amended to consider losses introduced by the optical antenna. For example, an uncoated lens will reflect approximately 4 percent of the radiation striking each of its surfaces. Similarly, an aluminum reflector may absorb 5 percent or more of near-infrared radiation striking it. A conservative accounting of these losses may be had by simply multiplying the result of Equation 2-4 by 0.9.

RECEIVER ANTENNA GAIN

Receiver antenna gain is simply the ratio of the collection area of a receiving antenna to the active surface area of the detector. For example, a receiver lens having a diameter of 10 centimeters gives a gain of 9637 when paired with a detector such as the EG&G SGD-040, which has an active area of 0.00815 cm^2. The gain figure can be reduced substantially if the lens is imperfectly focused or if the transmitter and receiver are not in good alignment.

Receiver antenna efficiency is the ratio of total radiation transmitted by a source to that collected by a receiver. In most cases, such a figure of merit is impractical because of variations in range from the transmitter to the receiver, atmospheric effects, etc. Therefore, for most practical purposes, receiver antenna gain can be expressed as:

$$G_{RA} = \frac{A_A}{D_A}$$

(Eq. 2-5)

where,

G_{RA} is the gain of the receiver antenna,
A_A is the collection area of the antenna,
D_A is the active area of the detector.

The same antenna losses of reflection and absorption that apply to transmitter antennas apply to receiver antennas. Therefore, a more conservative measure of receiver antenna gain may be had by multiplying the result of Equation 2-5 by 0.9.

COMMUNICATIONS-RANGE EQUATION

An experimenter can assemble simple light-beam communicators from a minimum of components even if he has little or no knowledge of how the system operates. For more-advanced communicators, however, a knowledge of the factors which determine maximum communication range can be very valuable.

The optical-communication-range equation permits the experimenter to predict the transmission range of a system without actually testing the unit in the field. In fact, the equation permits a communicator to be designed, tested, and evaluated *on paper* without the experimenter having constructed a single circuit.

The derivation of the range equation is straightforward and provides valuable insight concerning its usefulness. Fig. 2-10 shows a beam of light having a circular cross section projected from a transmitter, T. At a distance, R, from the transmitter, a receiver aperture having a diameter, d, intercepts some of the radiation from T. At the receiver, the diameter of the beam from T is D.

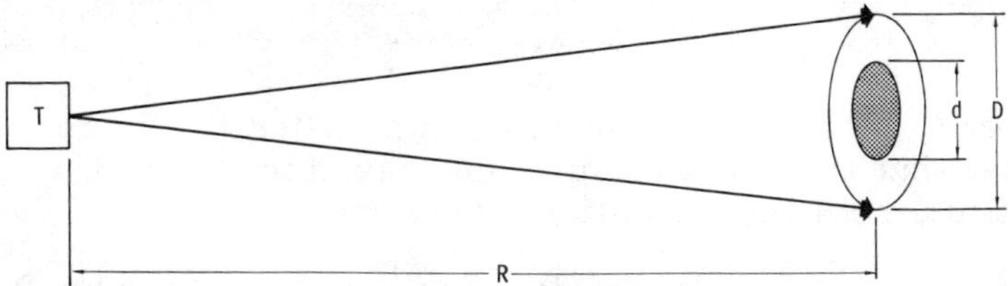

Fig. 2-10. Ratio of receiver aperture d to beam spot size D at range R.

To generate an equation defining the range of an optical communicator, we must first determine the signal level received by an aperture located some distance from the transmitter. Referring to Fig. 2-10, the signal intercepted by the receiver aperture is obviously the ratio of the area of the receiver to the area of the entire beam, less any losses contributed by the passage of the beam through the atmosphere and the receiver optics. We then have:

$$P_{rec} = (P_o) \left(\frac{A_{rec}}{A_{XMTR\ BEAM}} \right) (\tau) \qquad \text{(Eq. 2-6)}$$

where,

P_{rec} is the received power,

P_o is the total power in the transmitted beam,

A_{rec} is the receiver-aperture area,

$A_{XMTR\ BEAM}$ is the beam cross-section area (at the receiver),

τ is the loss factor contributed by the atmosphere and the receiver optics.

From Equation 2-2, we know that the diameter of the illuminated field at a range, R, from a source is equal to R times the divergence of the beam in radians. We know that the area of a circle is:

$$A = \pi r^2 \qquad \text{(Eq. 2-7)}$$

or,

$$A = \pi \frac{d^2}{4} \qquad \text{(Eq. 2-8)}$$

where,

r is the radius of the circle,

d is the diameter of the circle.

Knowing this, we can define the receiver and the transmitted-beam areas. The receiver area is:

$$A_{rec} = \pi r^2. \qquad \text{(Eq. 2-9)}$$

And the transmitted-beam cross-sectional area is:

$$A_{\text{XMTR BEAM}} = \frac{\pi r^2 \theta^2}{4} \qquad \text{(Eq. 2-10)}$$

where R is the range from the transmitter to the receiver.

Now that the two area functions have been defined, we can substitute both into Equation 2-6 to get:

$$P_{rec} = (P_o\tau) \left(\frac{\pi r^2}{\dfrac{\pi R^2 \theta^2}{4}} \right) \qquad \text{(Eq. 2-11)}$$

which reduces to

$$P_{rec} = \frac{4 P_o A_{rec} \tau}{\pi R^2 \theta^2} \qquad \text{(Eq. 2-12)}$$

Sometimes Equation 2-12 is condensed even further by cancelling 4 and π. This merely gives a more conservative result since $4/\pi$ is 1.27. Cancelling 4 and π gives a result having 78 percent of the actual value. Because of inhomogeneities in the power distribution of the transmitted beam, even Equation 2-12 gives a very conservative result for most cases. So, for more realistic results it is best not to cancel 4 and π.

Our derivation of the basic optical-communication-range equation is basically complete, but several additional considerations can make the equation more useful. For example, the equation can be used to solve for the maximum transmission range, R:

$$R = \sqrt{\frac{4 P_0 A_{rec} \tau}{\pi P_{th} \theta^2}} \qquad \text{(Eq. 2-13)}$$

Note that Equation 2-13 illustrates the inverse square law which states that the intensity of a diverging light beam varies as the square of the distance from the source. Notice also that in solving for R, P_{rec} became P_{th}. We have assumed a minimum level of received power necessary to trigger an output from the receiver. Therefore, P_{th} is the power threshold of the receiver.

APPLYING THE RANGE EQUATION

Because of inhomogeneities in the power distribution of a light beam, application of the range equation may give misleading results. For example, the power density of a typical light beam (Fig. 2-11) is normally higher near the axis than at the edge.

Equations can be and have been derived to predict the power density at any point in the cross section of a projected beam having a regular power distribution. These equations incorporate relatively complex statistical relationships and typically apply to beams having a normal or Gaussian power distribution. While such equations can be very useful in predicting the performance of a light-beam communicator, they do not necessarily consider subtle variations in the beam-power distribution. For this reason, an even more accurate way to predict the range of an optical communicator is to perform a relatively simple laboratory test with a model that incorporates at least the optical source and the optical-collimation system of the proposed communicator.

The test is performed by first assigning a maximum pointing error. This error angle then becomes the transmitter divergence which is fed into the range equation (Equation 2-13). The transmitter output then becomes only that power contained within the error angle. This power can be estimated, but best results are had by actually measuring the power within the error angle. This can be done by means of a radiometer and an aperture that blocks radiation outside the error angle. Fig. 2-12 is a photograph of a commercial radiometer which can be used for this purpose.

This method of applying the range equation gives extraordinarily good results so long as a proper accounting of atmospheric effects is made.

RECEIVER THRESHOLD LEVEL

The receiver threshold level in a pulse-modulated communication system is important in determining the system's signal-

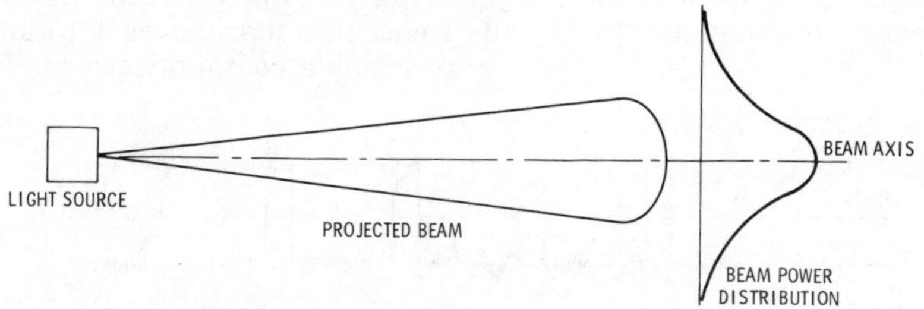

Fig. 2-11. Power density in a typical light beam.

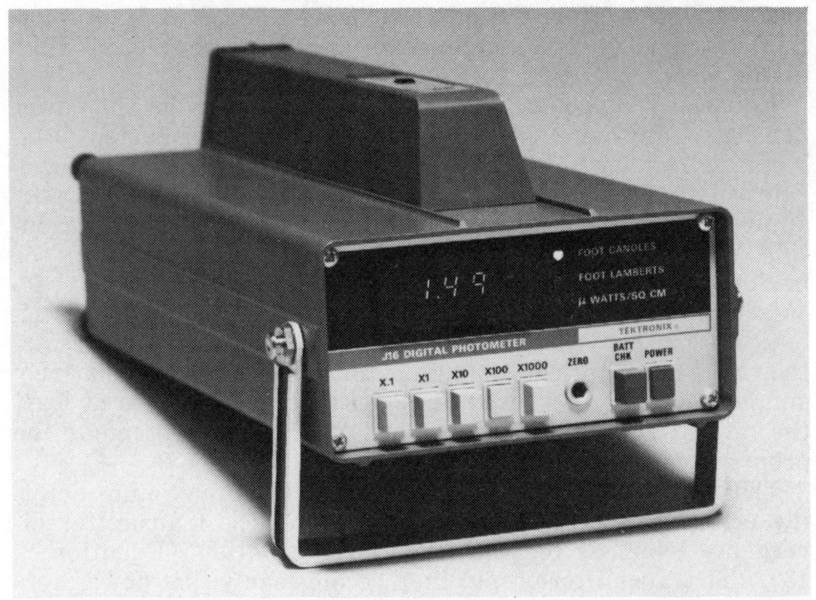

Courtesy Tektronix, Inc.

Fig. 2-12. Commercial radiometer/photometer with digital readout.

to-noise ratio (snr). Fig. 2-13 shows a typical pulsed signal that has been detected and amplified by an optical receiver. The receiver noise level is also shown. The threshold must be set so that consistent reception occurs over a wide range of conditions. A variable-threshold control can be included to permit optimizing the system for unexpected atmospheric conditions, background illumination variations, etc.

Engineers studying light-beam communications at RCA have shown that the optimum snr at threshold for one kind of laser diode pulsed-communication system is 32, or 15 dB. Since studies conducted at RCA and elsewhere reveal that the received signal level of a light beam passing through the atmosphere fluctuates 10 dB, and sometimes more, over a 10-km path, the 15-dB snr helps ensure reliable communications. If

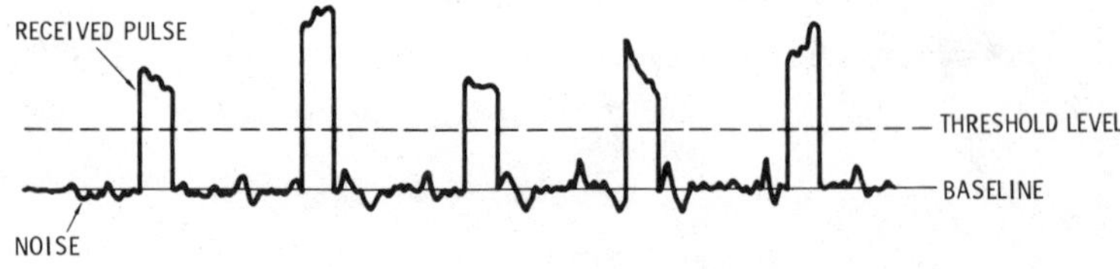

Fig. 2-13. Use of threshold level in pulse-modulated receiver.

52

occasional noise in the received signal can be tolerated, the receiver threshold level can be lowered for increased range capability.

BACKGROUND ILLUMINATION

Illumination from the sun, moon, stars, and artificial sources can have an adverse effect on the performance of an optical-communications receiver and can make application of the range equation difficult. All these radiant sources can add noise to the receiver detector and amplifier, and this noise serves to reduce the snr.

The intercepted background power will vary at different times of the day and year, at different locations, and for various backgrounds. In the latter case, a sunlit cloud will contribute considerably more background power than blue sky.

The adverse effects of background illumination can be reduced by decreasing the receiver beam width (field of view), operating at night, and by avoiding backgrounds with high reflectance. The use of an optical filter can also reduce the problem. In an infrared communication system, for example, a spike filter that passes only the infrared radiated by the transmitter may significantly improve daylight operation of a receiver.

It should not be assumed that an optical filter always improves the performance of an optical receiver, since the filter itself does not transmit 100 percent of the beam. Absorption filters may transmit only 50 to 70 percent of the beam, while interference filters may transmit as much as 90 percent of the beam.

CHAPTER 3

Light Sources

Many kinds of light sources have been utilized in both experimental and operational light-beam communicators. Some sources are best suited for analog modulation, others for pulse modulation, and still others are suited for both modulation formats. Many sources can be directly modulated, while others must be modulated by external means.

In this chapter the major light sources for optical communicators will be described. Although a source such as the sun may seem to be of historical interest only, voice-modulated sunlight communicators are still being experimented with. Most of this chapter is devoted to semiconductor light sources and lasers since these comparatively recent sources offer the best potential for practicable optical communications.

THE SUN

Alexander Graham Bell used reflected sunlight in his historic experiments with the Photophone in 1880. Sunlight is appealing for use in a communication system since it is both abundant and free. For this reason, as late as World War II some military voice communicators, such as some of the German *Lichtsprechers,* used reflected sunlight for communicating.

Bell's original Photophone modulated a beam of sunlight by means of a thin, flexible mirror directly modulated by the voice of the speaker. Fig. 3-1 is a photograph of a young lady using a modern version of the Photophone. The transmitter consists of an aluminum cylinder to which a thin mirror is cemented. The receiver is a silicon solar cell mounted in the reflector of a

Fig. 3-1. Christina Schneider demonstrating a simple sunlight communicator.

commercial flashlight. A transistor amplifier boosts signals from the solar cells in order to operate a small, self-contained speaker. The apparatus shown in Fig. 3-1 has a range of several hundred feet.

Fig. 2-14 in the previous chapter shows that the spectral distribution of sunlight is so broad that numerous kinds of detectors can be used in a sunlight link. Bell and many others used selenium, and selenium cells are available today which will operate in such a role. Best results, however, are had with silicon detectors. Large area detectors, such as solar cells, may not require an external lens or reflector, while small area detectors, such as phototransistors, definitely do.

The effectiveness of a sunlight communicator is limited by the availability of sunlight. Furthermore, sunlight illuminance varies from 12.4×10^4 lumens per square meter when the sun is directly overhead to 1.09×10^4 lumens per square meter when the solar angle is 10 degrees. Clouds and even haze prevent sunlight communications.

Sunlight communicators must be designed to compensate for the rotation of the earth on its axis. This is not a major problem for short-range, hand-held units such as the one shown in Fig. 3-1.

Finally, the natural beam divergence of reflected sunlight is the angle the sun itself subtends or 9 milliradians (0.5 degree). Suitable optical systems can be employed to reduce or increase this angle or to concentrate more light onto a modulator.

INCANDESCENT LAMPS

Incandescent lamps have been used in both directly and externally modulated light-beam communicators. No special requirements are imposed on lamps designed to be modulated externally, but directly modulated lamps must possess a filament having relatively small mass.

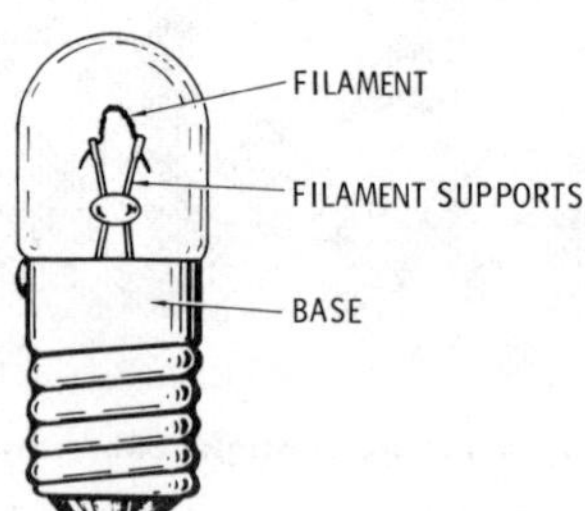

Fig. 3-2. Construction of typical tungsten lamp.

A typical tungsten lamp consists of an evacuated glass envelope and a metal base. Filament supports attached to the base, as shown in Fig. 3-2, support the filament near the center of the envelope.

Nearly all incandescent lamps incorporate a tungsten, or sometimes a tungsten-rhenium alloy, filament. Tungsten is a high-refractory metal and can be heated to a high degree of incandescence before its evaporation rate becomes significant. Since tungsten has a relatively low electrical resistance, it is ideal for use as a lamp filament.

Several construction methods are employed in the manufacture of tungsten lamp filaments, and the three most common are the single strand, the single coil, and the coiled coil. The filament design can be important in an incandescent-lamp optical communicator since a well-collimated filament lamp will produce an image of the filament at the receiver. If the filament is a coil, the power distribution will be nonuniform and may cause reception problems. This effect is illustrated in Fig. 3-3.

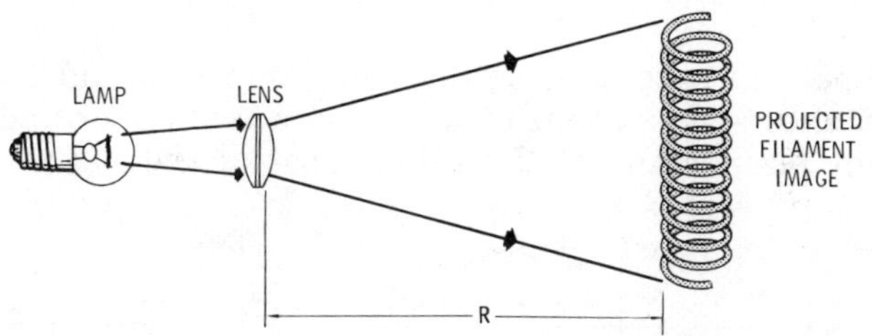

Fig. 3-3. Filament image projected by focused incandescent source.

Since nearly all incandescent lamps have a spherical power distribution, their radiations can be effectively collected and collimated with a parabolic reflector. A common flashlight is an excellent example of an incandescent lamp mounted within a parabolic reflector.

A consideration of radiant rise time (the time a filament requires to attain 90 percent of full brightness) is important when you are choosing a lamp for direct modulation. Usually, lamps having low-mass filaments have reasonably fast rise times, and the rise time of some miniature lamps may be 10 milliseconds.

Of course a rise time of even 10 milliseconds indicates a 100-percent modulation capability of only 100 Hz at best, so audio frequencies of a few thousand hertz can be utilized only by prebiasing the lamp to some average brightness level and by modulating at well under 100 percent.

Several direct-modulation techniques have been employed with incandescent lamps, and one of the most popular is shown in Fig. 3-4. This circuit can be directly connected to the output transformer of many kinds of radios and audio amplifiers. The battery prebiases the lamp to some average bias, and the resistor limits the lamp current to a safe value to prevent the filament from being burned out at high-modulation levels. The author has used the circuit in Fig. 3-4 to send audio signals several thousand feet to a simple receiver consisting of a silicon solar cell and an audio amplifier.

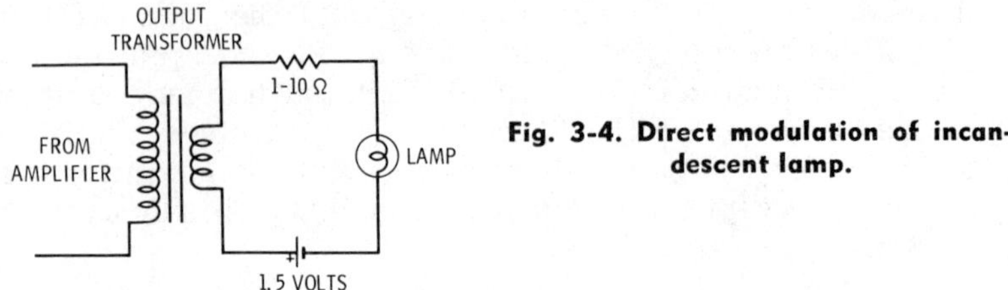

Fig. 3-4. Direct modulation of incandescent lamp.

The spectral output of a tungsten lamp is broad band, and Fig. 3-5 shows the output of a tungsten lamp operated at several color temperatures. Note that most of the radiation from

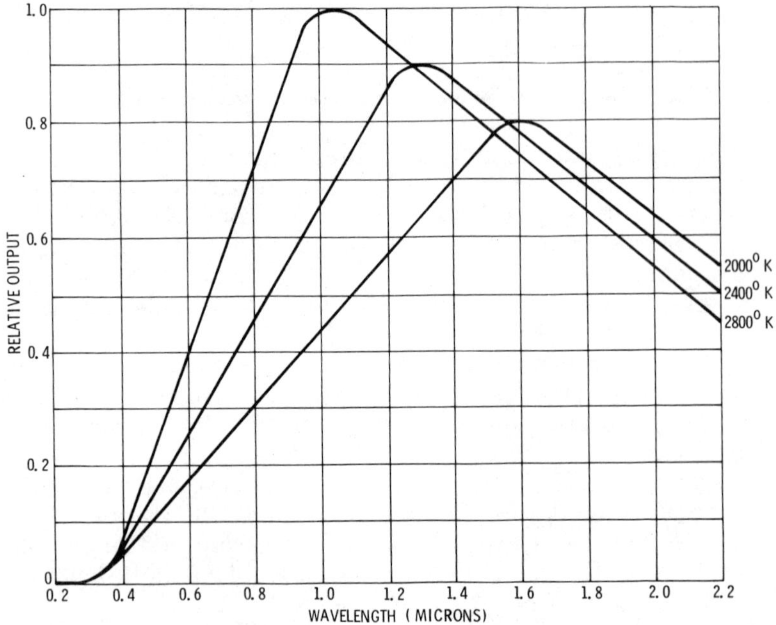

Fig. 3-5. Tungsten lamp spectral output versus filament color temperature.

a tungsten filament is infrared. The broad-band nature of a tungsten source means many different detectors can be employed in a receiver system.

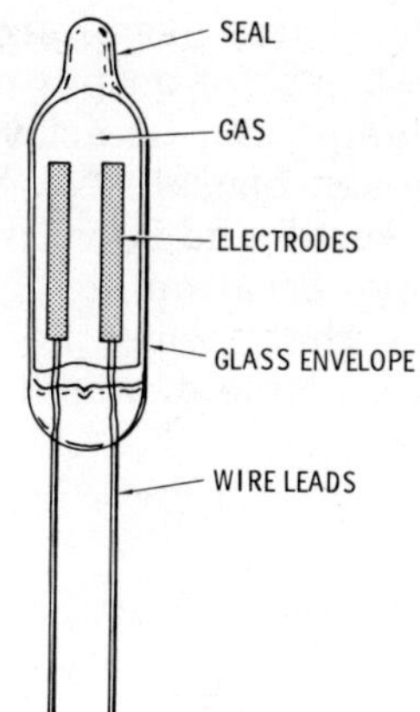

Fig. 3-6. Glow-lamp construction.

GLOW-DISCHARGE LAMPS

The glow lamp shown in Fig. 3-6 consists of a glass envelope containing two or more electrodes and filled with a gas capable of being ionized. The most common glow lamps contain neon or argon. Neon produces a yellow-red range of wavelengths

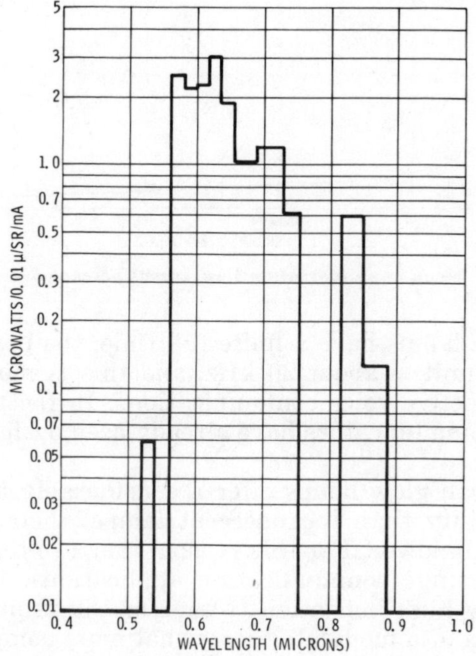

Fig. 3-7. Spectral output of a neon lamp.

when ionized, and argon produces ultraviolet and blue. Fig. 3-7 shows the spectral output of a typical neon lamp.

Neon lamps can be activated within a few hundred microseconds when operated at their normal turn-on voltage (usually between 65 and 90 volts). Much faster ionization times are possible by operation at higher than normal voltages, and Fig. 3-8 shows that twice the ionization, or breakdown voltage, yields a rise time of about 10 microseconds.

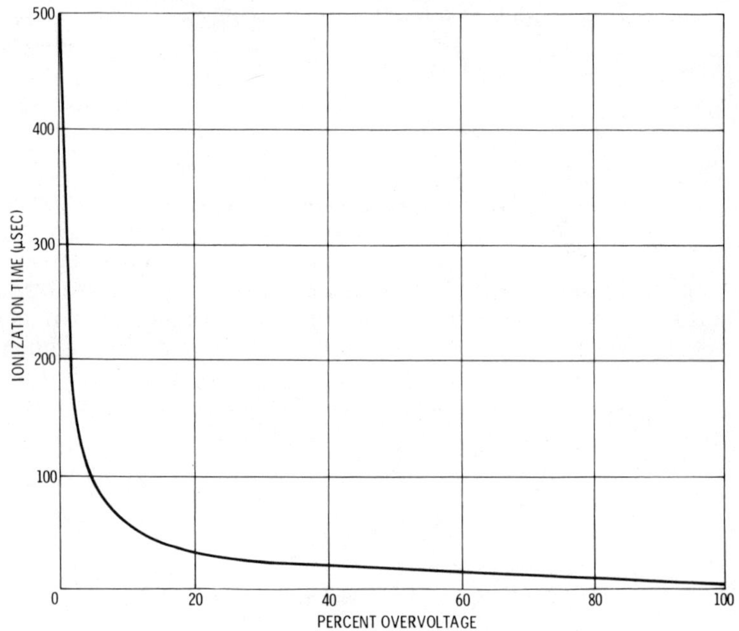

Fig. 3-8. Neon lamp ionization time versus percent overvoltage.

Since neon lamps have a finite fall time, the practical upper modulation limit is about 20 kHz, and this is more than adequate for effective voice communications. In fact, a variety of glow-lamp communicators have already been designed and constructed.

Though neon glow lamps offer a considerably higher modulation capability than incandescent lamps, their light output is significantly lower. For this reason, they are limited to relatively short-range communication applications. Furthermore, the relatively high ionization voltage of the neon lamp makes circuit design of a modulator somewhat more complex than the simple format for an incandescent lamp presented in Fig. 3-4.

ARC LAMPS

Since arc lamps operate at considerably higher temperatures than most light sources, they are among the most brilliant artificial-light sources. One of the earliest arc lamps consists of two carbon rods separated by an air space. When an arc is established between the two rods by means of a suitable dc voltage, the positive carbon, which contains a core of material softer than the carbon shell, becomes white hot. Since the core of the positive rod is softer than its carbon exterior, it burns away at a slightly faster rate and forms a cavity within the rod. The cavity serves to confine the gases generated by the burning material, and these gases contribute to the output of the carbon arc lamp.

Carbon arc lamps have found extensive use in military searchlights, theatrical spotlights, and motion-picture projectors. In recent years, however, the xenon arc lamp and other high-intensity sources have begun to replace carbon arcs.

Two very useful arc lamps are the tungsten arc and the zirconium arc. Both these lamps consist of a central tubular cathode and a flat anode containing a central hole. The arc is established between the end of the cathode and the anode. Fig. 3-9 shows the construction of a typical zirconium arc lamp.

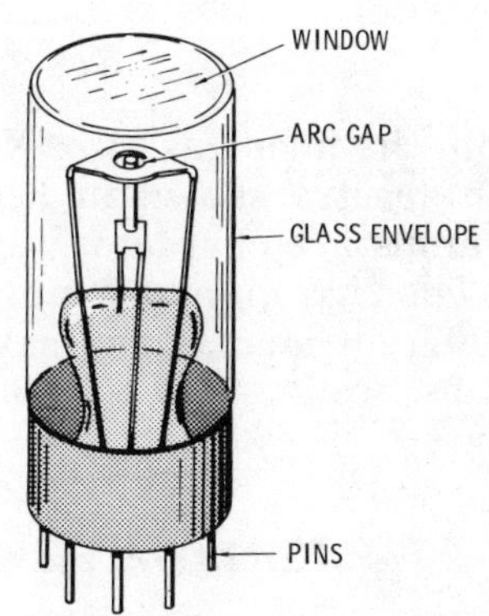

Fig. 3-9. Zirconium arc lamp construction.

The zirconium arc lamp produces a whiter light than the tungsten arc lamp. The tubular cathode is packed with zirconium oxide which reaches a temperature of 2700 degrees Celsius, sufficiently hot to produce nearly as much radiation as a carbon arc lamp.

Fig. 3-10 shows a short-arc xenon arc lamp of the type invented by Paul Schulz of Germany during and after World

War II. This lamp consists of a heavy-walled quartz bulb filled with xenon at a pressure of 20 to 40 atmospheres. The electrodes, which are large in order to dissipate heat, are made of tungsten. While low-pressure arc lamps produce light by electrode heating, high-pressure arcs produce a brilliant emis-

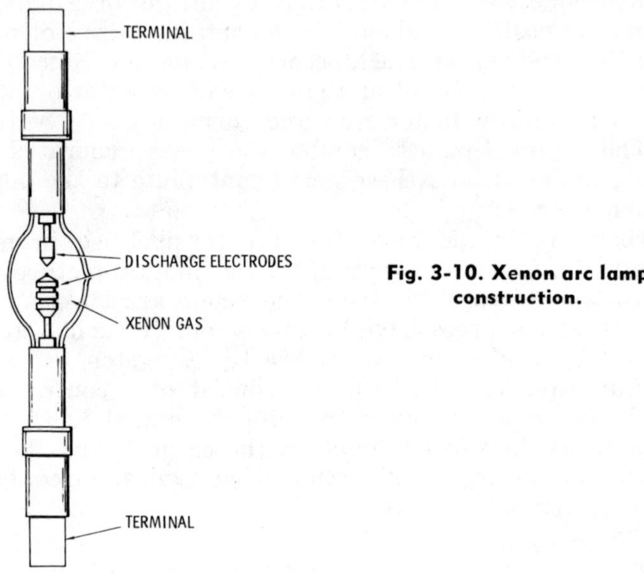

Fig. 3-10. Xenon arc lamp construction.

sion from the filler gas also. Xenon, for example, produces the emission spectra shown in Fig. 3-11.

Arc lamps are characterized by a very small light-emitting region. The zirconium arc lamp, for example, may have an arc only 0.007 cm long. Their small light-emitting regions make arc lamps well suited for use in narrow-beam projection systems.

LIGHT EMITTING DIODES (LEDs)

Some types of semiconductors can be caused to generate efficient visible or near-infrared radiation when they are given a pn junction and stimulated with an electrical current. The light-emission process in a junction diode is summarized in Fig. 3-12 in a graph called an *energy level diagram*. The valence and conduction bands on each side of the pn junction are connected by sloping lines representing the junction's barrier to electron flow. When electrons are injected into the n side

62

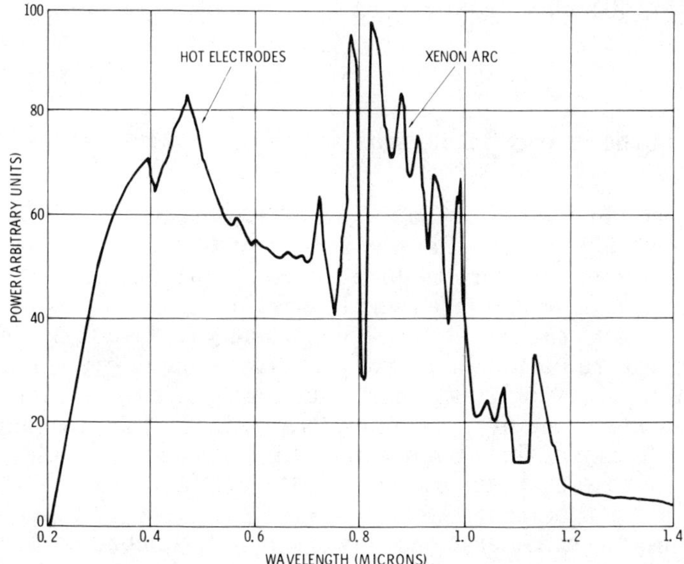

Fig. 3-11. Typical xenon arc lamp spectrum.

by a current source, the junction's potential barrier is reduced and current flow occurs.

The radiation wavelengths emitted by a semiconductor junction are related to the energy difference, or *band gap,* between the valence and conduction bands of the semiconductor. For a distinct band-gap energy, the following formula applies:

$$\tau = \frac{hc}{E} \qquad \text{(Eq. 3-1)}$$

where,

τ is the radiated wavelength,
h is Planck's constant (6.63×10^{-34} joule-seconds),
c is the velocity of light (3×10^{14} micrometers per second),
E is the energy in joules that separates the valence and the conduction bands.

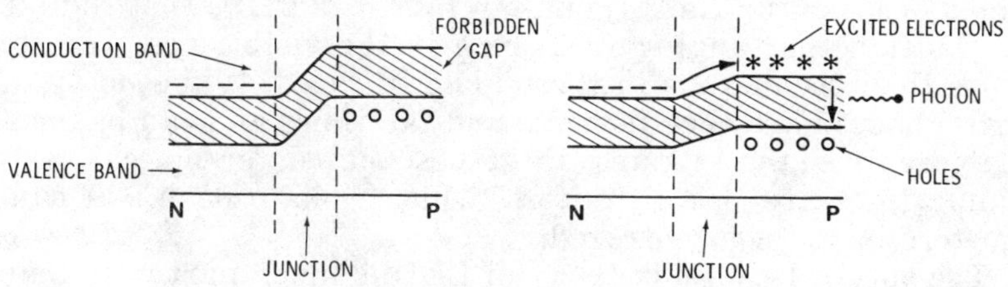

Fig. 3-12. Energy-level diagram of a pn junction.

Equation 3-1 may be simplified to:

$$\tau = \frac{1237 \text{ nanometers}}{E_g} \qquad \text{(Eq. 3-2)}$$

where,

Eg is the energy in electron-volts between the valence and
the conduction bands.

A semiconductor with a band gap of 1.36 electron-volts would
then emit photons with a wavelength of 909 nanometers.

Most semiconductors require the controlled addition of dop-
ant impurities to increase their electron-hole concentration and
mobility, and the result may be a variety of band-gap levels
that cause radiation of a lower energy (longer wavelength)
than that of the original semiconductor. This can be beneficial
since semiconductors are more transparent to wavelengths
having less than their band-gap energy. This means solid-state
light sources that absorb very little of their recombination
radiation can be fabricated.

Band-gap considerations make some semiconductors far
more efficient radiant sources than others. For example, both
silicon and germanium pn-junction diodes emit a small amount
of near-infrared radiation when forward biased; but, since
silicon and germanium have indirect band gaps, the radiant
emission is inefficient. The indirect band gap means intermediate energy levels are present which give rise to heat radiation at a much higher efficiency.

Gallium arsenide (GaAs), a compound semiconductor, has
a direct band gap and is therefore a very efficient optical
source. In 1965, William N. Carr of Texas Instruments measured an internal quantum efficiency of 88 percent in specially
prepared GaAs diodes. This means 88 percent of the electrons
injected into the diode gave rise to the emission of a photon.
Because of absorption within the GaAs crystal, the diodes exhibited an external quantum efficiency of only 42 percent—
which is still quite high.

By utilizing various semiconductor compounds and alloys
(many containing GaAs), one can fabricate LEDs that radiate
at wavelengths ranging from blue in the visible spectrum to
more than 30 microns in the middle infrared. While many of
these diodes have very fast rise and fall times and can be modulated at megahertz rates, those best suited for optical communications are GaAs emitters, which radiate near 900 nanometers in the near infrared.

The spectral emission from an LED is much narrower than
that from most other light sources except the laser. For ex-

ample, the emission from a typical GaAs LED may be 25 nanometers wide. This narrow spectral output permits the use of a narrow-band optical filter at the receiver to block unwanted optical signals. Fig. 3-13 shows the spectral output of several LEDs.

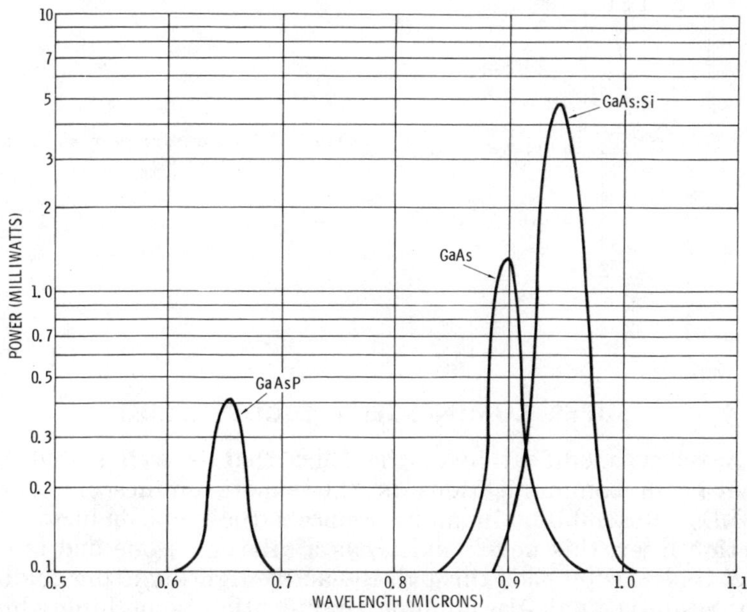

Fig. 3-13. Spectral and typical power output of several LED semiconductors.

Fig. 3-14 shows the construction of a typical LED. The metal heat sink, which serves to absorb excess heat generated within the diode, can also reflect photons which might otherwise be wasted.

Since GaAs and other LED compounds have a high index of refraction, light striking the surface of such materials at an angle of more than approximately 17 degrees is reflected back into the crystal. Several different construction methods are employed to reduce this optical loss. In one, the LED chip is ground into a hemispherical dome. The rounded surface of the dome means that light originating at a centrally located pn junction will always strike the crystal surface at less than a 17-degree angle. A less expensive light-extraction technique is to cover the LED chip with a transparent epoxy having an index of refraction between that of the crystal and the air. This permits more of the light to be collected from the chip.

A detailed description of LED design, construction and operation is given in *Light Emitting Diodes* (by Forrest M. Mims, Howard W. Sams & Co., Inc., 1973).

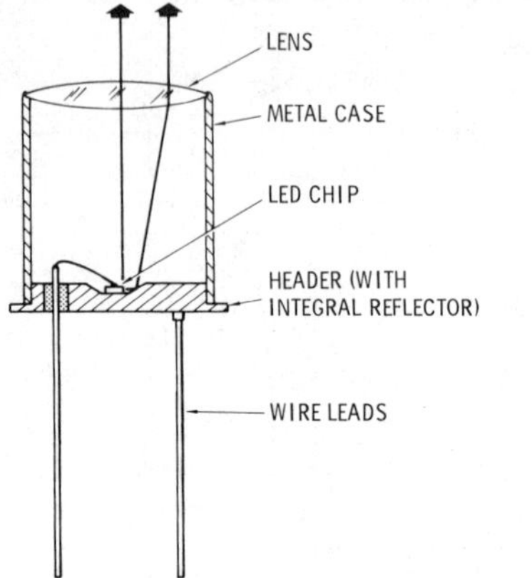

Fig. 3-14. Construction of a typical LED.

SUPER LUMINESCENT DIODES (SLDs)

A recent modification of the LED that is well suited for light-beam communications is the super luminescent diode (SLD). Resembling in many respects the semiconductor injection laser, this novel device has a mirror on one end to reflect its emission back through the active light-emitting region. The result is single-pass photon amplification, something which does not occur in a conventional LED.

The SLD produces a much narrower beam than any other LED. This means the output radiation of the SLD can be coupled with much higher efficiency into an optical fiber than can most LEDs and even some kinds of semiconductor lasers.

LASERS

The optical source most responsible for the current rush to develop practical light-beam communication systems is the laser. The first laser was operated by Theodore Maiman in 1960 at the Hughes Aircraft Company. Since then, many kinds of lasers have been devised, some of which are particularly well suited for optical communications.

The word *laser* is an acronym for *L*ight *A*mplification by *S*timulated *E*mission of *R*adiation. A prerequisite for laser action is a substance capable of being excited to a higher than normal energy level by an external energy source. If more atom molecules or ions in the substance are in an excited rather than unexcited state (*population inversion*), and if the active material is provided with an optically resonant cavity, laser action will occur.

Maiman's original laser utilized a ruby rod for the active material. Ruby consists of aluminum oxide (Al_2O_3) doped with a small quantity of chromium. The chromium ions occupy aluminum sites in the Al_2O_3 matrix and provide the optically active properties of ruby.

The ruby rod was provided with an optically resonant cavity by polishing both its ends until they were perfectly parallel and flat. The external energy was supplied by a coiled xenon flash tube placed over the rod.

The lasing process was initiated when a large capacitor was discharged through the flash lamp. Some of the flash-lamp radiation was absorbed by the ruby and many of the chromium atoms were excited to a higher than normal energy level. When a population inversion of excited chromium atoms was reached, photons spontaneously emitted by some excited atoms begin to trigger, or stimulate, the emission of large

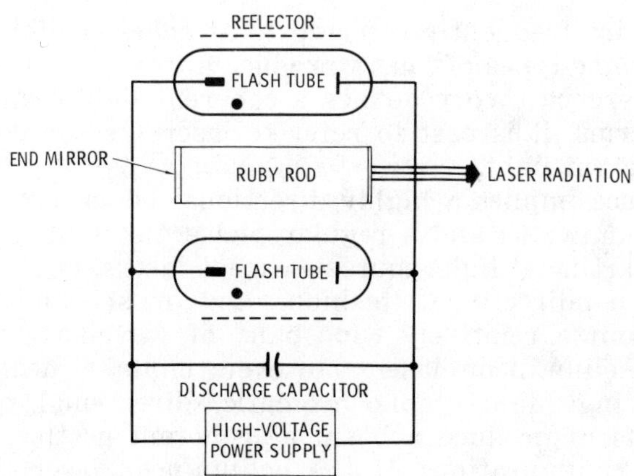

Fig. 3-15. Operating diagram of a ruby laser.

numbers of photons from other excited atoms. The stimulation process was reinforced by two silver coatings placed on each end of the ruby rod. These mirrors, which formed the resonant cavity, caused photons to bounce back into the ruby where they stimulated additional emission. The result was a standing wave of light which eventually emerged from one of the mirrors. The light emerged in a series of very narrow, powerful pulses lasting a total of several milliseconds. The mirror through which the light emerged was partially transparent; the other mirror was nontransparent.

Fig. 3-15 shows the construction of a ruby laser. The radiation emitted by Maiman's ruby laser was a brilliant red at a wavelength of 694.3 nm. Although ruby remains an important laser material, since 1960 many combinations of active materials, resonant cavities, and excitation methods have been utilized to produce laser action. In fact, literally hundreds of solids, liquids, and gases have been used to produce thousands of laser frequencies at wavelengths ranging from the ultraviolet to the far infrared.

Since the laser is so important to the future of optical communications, the remainder of this chapter will consider several categories of lasers. Those particularly well suited to optical communications will be described in detail. First, though, let us digress for a few pages with a discussion of laser light since it is the unique nature of radiation emitted by a laser that makes communications such an important application.

LASER LIGHT

Lasers are frequently referred to as *coherent* light sources, but only some types of lasers produce a totally coherent beam. Since coherence theory defines a coherent light beam in very precise terms, it is best to refer to lasers as sources of partially coherent light.

Coherence implies a highly directional beam having a very narrow bandwidth and a regular phase distribution. All natural and artificial light sources, except lasers, emit light in a random, nondirectional fashion, and most nonlaser light sources emit a relatively wide band of wavelengths. Even a low-cost helium-neon laser, however, emits a near-coherent beam having highly monochromatic, directional properties. Not all lasers produce a beam as coherent as that produced by many configurations of the helium-neon laser. Nevertheless, the beam from *any* laser is characterized by one or more of the following properties:

1. Monochromaticity—The spectral output from most lasers consists of a narrow spectral packet containing very narrow, individual frequency spikes. Thus, a semiconductor laser may emit a band of wavelengths having a total width of perhaps 2 nanometers and containing individual frequency spikes only 0.01 nm wide. Carefully fabricated gas lasers can emit considerably purer beams. For example, a properly adjusted helium-neon laser can emit a single wavelength having a width of less than 10^{-8} nm. The monochromatic nature of a laser beam is sometimes referred to as *temporal* or *spectral coherence.*

The monochromatic nature of laser light is important to laser-communication systems. Since the spectral width is so narrow, optical passband filters can be used to block unwanted light from a laser receiver while passing only the laser wavelength. Additionally, the narrow spectral bandwidth of most lasers is directly related to the ultimate capacity of a laser communications link.

2. Directionality—The beam divergence of a laser operating in the lowest-order transverse-electromagnetic mode is defined by the size of the laser's output aperture and the adverse effects of the atmosphere. In a vacuum, the beam from such a laser would be completely parallel with no angular divergence. But, diffraction caused by the slight bending of light rays at the laser's output aperture eventually causes the beam to spread slightly.

Most lasers do not normally operate in the single lowest-order mode, and since different modes are emitted at slightly different angles, most lasers do not emit perfectly parallel beams. Nevertheless, the beam emitted by most lasers is far narrower than that emitted by conventional light sources. This directional property of laser light is sometimes referred to as *spatial coherence.*

The directional nature of laser light is important for optical communications and numerous other applications. All of the narrow beam emitted by a laser is easily collected by simple optics for reprojection as a wider or even narrower beam. This gives very high efficiency to a light-beam communication system. The beam from some lasers is sufficiently narrow that no external optics are required.

3. Phase—By means of a filter, a pinhole aperture, and a collimating lens, a conventional light source can be caused to emit a beam having some of the monochromatic and directional properties of a laser. However, the power density of the conventional light would be significantly less than that of even a low-powered laser. Furthermore, the amplitude of the filtered conventional source would exhibit distinct amplitude variations due to the random nature of light generated by a nonlaser source. The output of a laser is stabilized in amplitude since the laser is a true sinusoidal oscillator having a uniform phase.

The stable-wave amplitude of the laser is important to various types of optical communication and modulation techniques that are employed.

4. Intensity and Brightness—The light output of even a very small laser is incredibly more intense and brilliant than that of any other light source. A. E. Siegman of Stanford University has calculated that the photon output of typical visible-wavelength lasers ranges from 10^{16} to 10^{28} per second, while that of a typical, equivalent-aperture thermal source is no more than 10^{12} in a wavelength range of 100 nm. Since the wavelength of even a simple gas laser may be less than 0.001 nm, the laser emits an incredibly intense beam.

The laser is also substantially brighter than any other light source. Siegman has calculated that a thermal light source filtered and collimated to have the same bandwidth and beam spread as a typical laser would have to have a temperature of 10^{16} degrees Kelvin. Therefore, the very narrow beam and frequency of a laser make it enormously brighter than any other light source.

Many of these properties of laser light were vividly illustrated to the author in a recent field test of a laser-diode voice communicator. Semiconductor lasers made from gallium arsenide emit a beam significantly less coherent than that emitted by other lasers. Yet at approximately 1 km on a bright sunny day, the beam from the laser diode was the brightest light source in view. Since the beam was infrared (905 nm) and invisible to the unaided eye, the laser was viewed through an image-converter tube which transformed near-infrared wavelengths into visible light. A narrow-bandwidth optical filter was placed over the image-tube aperture to permit only wavelengths at or near the laser wavelength to be passed to the tube's photocathode. The laser was operated just above threshold, yet it appeared far brighter than sunlight reflections from shiny metal objects nearby.

Fig. 3-16 sums up the characteristics of laser light as compared to those of light emitted by a conventional source.

LASER-MODULATION BANDWIDTH

Laser light is well suited for carrying enormous quantities of information because of its coherence properties and enormous frequency range. The frequency range of all radio and microwave channels totals less than 200,000 MHz, but the frequency range of the optical spectrum ranges from 50,000,000 MHz to 5,000,000,000 MHz! Obviously such a frequency range is capable of carrying enormous quantities of information.

For example, the widest-bandwidth laser communicator being considered as this is written will be modulated with pulses having a width of 30 picoseconds for a total bandwidth of 30 gigahertz. This one system alone exceeds the total rf and microwave spectrum and utilizes only 10 percent of the ultimate bandwidth in a laser beam with a frequency of 3×10^{14} Hz.

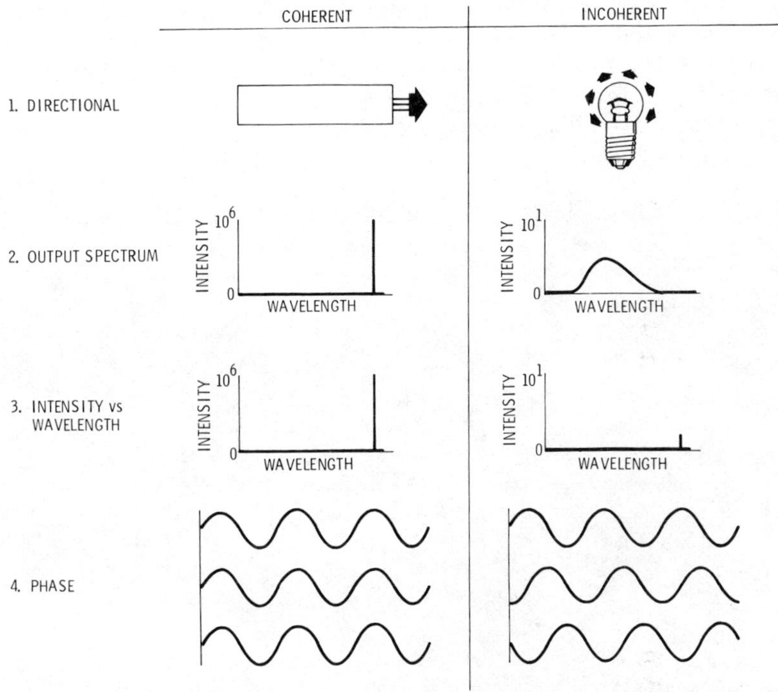

Fig. 3-16. Coherent versus incoherent light.

Only 1 percent of the optical spectrum has the theoretical capability of carrying some 375,000 8-MHz television broadcasts or nearly a billion 3.5-kHz audio channels. These astounding information rates are only theoretical, and numerous laboratories are working on methods of modulating and demodulating laser light in order to exploit even a small fraction of the available bandwidth.

GAS LASERS

Gas lasers are by far the most varied category of lasers. They can be excited into the laser state in more ways than any other laser, and they produce the widest range of wavelengths (235.8 nm in the ultraviolet to 774 microns in the infrared). Gas lasers include the most efficient and the most powerful of all lasers.

While many gas lasers are suitable for optical communications, the three best-studied systems are the helium-neon, argon, and carbon dioxide lasers. Scientists at Bell Laboratories achieved laser action with helium neon only months after Maiman operated the first laser. Helium-neon can emit infrared wavelengths, but most commercial HeNe lasers are designed to emit a bright red beam at 632.8 nm. Fig. 3-17 illustrates the very narrow beam produced by even a relatively inexpensive HeNe laser.

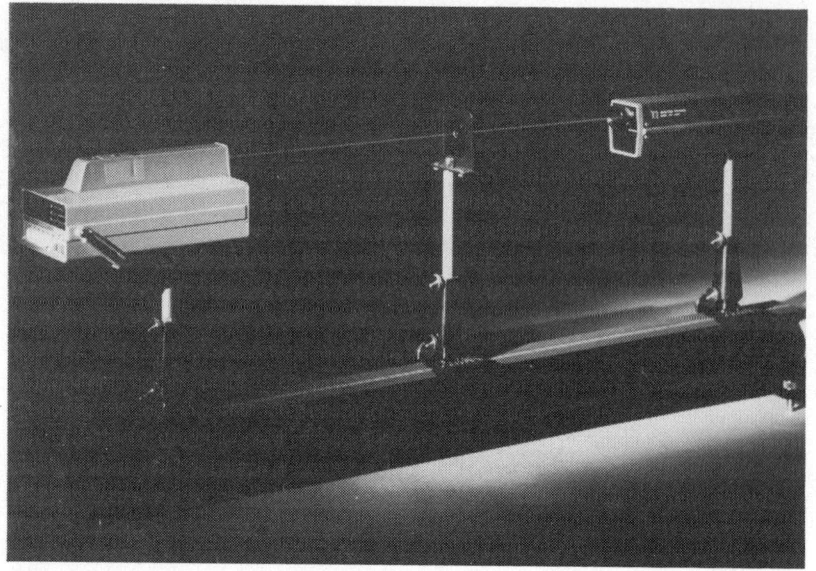

Courtesy Tektronix, Inc.

Fig. 3-17. The narrow beam of a HeNe gas laser entering a radiometer.

HeNe lasers produce less power than many other gas laser systems, but their level of development is so advanced that self-contained glass or glass-metal laser tubes are available for less than $50. Fig. 3-18 is a photograph showing an early HeNe laser tube. More-recent tubes use a sturdier coaxial design. These laser tubes contain discharge electrodes, a reservoir of helium and neon, and the two mirrors which form the optically resonant cavity. The mirrors are permanently aligned at the factory.

A simple HeNe laser tube requires only a high-voltage power supply to initiate operation and a lower-voltage power

72

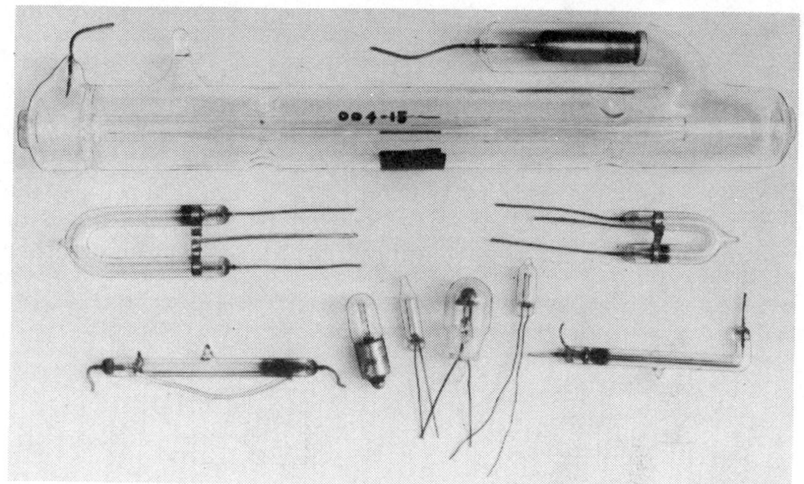

Fig. 3-18. Helium-neon laser tube (top) and assortment of neon, argon, and xenon lamps.

supply to continue operation. The operating life of the tube may exceed 20,000 hours of continuous operation. Since the cost of materials in such a tube is under $1, one major manufacturer predicts the price of individual laser tubes will eventually drop to about $5.

Due to their advanced stage of development. HeNe lasers have been used in a variety of audio- and video-communication experiments. Several manufacturers offer commercial HeNe laser-communication systems, and some of these are described in Chapter 6.

The argon laser is far more complex than the relatively simple HeNe device. Argon lasers require higher operating current densities, careful discharge-tube construction, and provision for removing the large amount of heat generated during operation.

Argon lasers are interesting to communication engineers since they can produce a highly coherent beam having considerably more power than HeNe units. Also, argon lasers are capable of producing 488- or 515-nm wavelengths in the blue and green portions of the spectrum, which are particularly well suited to transmission through water. For this reason, the Navy has conducted experiments in the underwater propagation of argon laser beams.

The carbon dioxide laser ranks among the most efficient of laser systems. Even a relatively unsophisticated CO_2 laser may have an efficiency of 20 or more percent. This compares to less than 0.1 percent for HeNe lasers and 0.2 percent for argon lasers.

Fig. 3-19 is a diagram of a simple CO_2 laser. This particular laser uses a flowing mixture of helium, nitrogen, and CO_2. Since the radiation produced by a CO_2 laser has a wavelength of 10.6 microns in the middle infrared, ordinary optical components cannot be used to transmit the beam. Therefore, the ends of the discharge tube are covered by windows made of sodium chloride (salt), germanium, gallium arsenide, or other exotic materials.

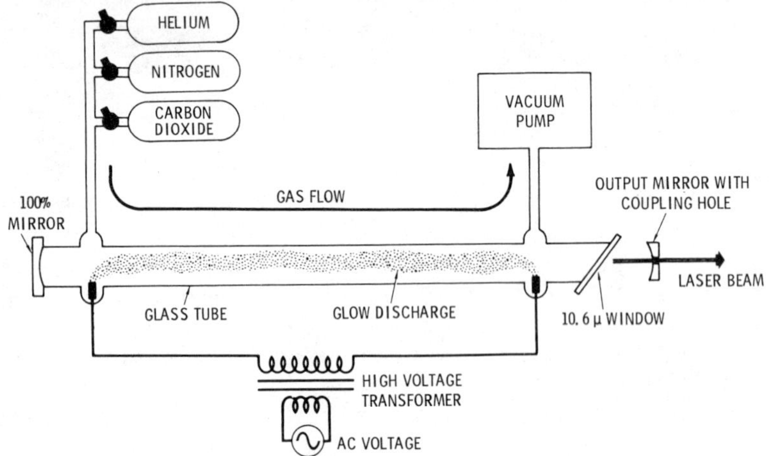

Fig. 3-19. Construction of a simple CO_2 laser.

Although both CO_2 and HeNe lasers can be very simple to construct, CO_2 systems produce considerably more power. Even a relatively small CO_2 laser can emit 100 watts or more, and CO_2 systems that continuously generate more than 10,000 watts have been constructed. Military research into laser weaponry based on CO_2 laser technology has resulted in classified laser systems which produce in excess of 100,000 watts.

The very high power output of the CO_2 laser means that special precautions must be taken during operation. Also, the cavity windows and mirrors must be capable of withstanding the very high power contained in the laser beam.

Since 10.6 microns falls in a high-transmission region of the atmosphere and since CO_2 lasers are very efficient radiant sources, communication engineers are particularly interested in the CO_2 laser for long-range space and atmospheric communications. A few problems remain to be solved before practical CO_2 lasers suited for communications are available. Better windows for 10.6-micron radiation and more-compact CO_2 lasers are under continual development. Additionally, high-data-rate modulation techniques and new kinds of efficient 10.6-micron detectors have been devised.

LIQUID LASERS

Three major classes of liquid lasers have been developed. The first uses a variety of fluorescent rare earths dissolved in a transparent solvent. The liquid is simply poured into a glass tube and excited by a flashlamp.

Another liquid laser employs neodymium dissolved in acid and placed in a tube. This laser is simple to operate, but the highly corrosive acid is very dangerous.

A third liquid laser utilizes organic dyes dissolved in a host liquid. Among the fastest-growing laser families, organic-dye lasers can be made to emit over the entire visible spectrum by choosing suitable dyes and end mirrors. Dye lasers are by far the most promising of liquid lasers.

Thus far, relatively little interest has been expressed in liquid lasers for communications. However, the fact that liquid lasers can be operated at higher power levels than many solid lasers simply by flowing the liquid through a cooling reservoir or radiator to remove excess heat, means that liquid lasers might ultimately find use in some laser-communication systems.

SOLID-STATE ION LASERS

Ruby was used in the first solid-state ion laser, and since then many ion-containing crystals and glasses have been made to lase. Ruby lasers find wide use in various areas of laser research, but their inefficiency and the difficulty of obtaining continuous operation make ruby generally unsuited for communication applications.

The most important solid-state ion laser for potential communications application is neodymium-doped yttrium-aluminum-garnet (YAG), which can be doped with other ions as well. The most prominent output for Nd:YAG is 1.06 microns in the near infrared.

YAG lasers have emitted up to 1000 watts in continuous operation. They are more efficient than HeNe and argon gas lasers, with measured power efficiencies ranging from 2.9 percent for a 100-watt laser unit to 1.7 percent for a 750-watt laser unit.

Since nonlinear optical crystals can be inserted into a YAG laser cavity to achieve second-harmonic frequency conversion, the invisible 1.06-micron output can be converted to a brilliant green at 530 nm (half of 1.06 microns). The fact that YAG lasers can be operated in a wide variety of configurations with relatively high efficiency makes them well suited for many optical-communication roles. In fact, YAG lasers are being considered for space, atmospheric, and optical-fiber communication links.

One of the most interesting YAG-laser systems employs semiconductor light emitting diodes to excite the YAG crystal. Most YAG lasers are excited by xenon flashlamps or by various types of continuous tungsten or arc lamps. These excitation methods are inefficient since the neodymium ion is excited only by certain wavelengths (called absorption bands). Certain LEDs emit efficient radiation at one of the major YAG absorption bands and can therefore be used as an efficient solid-state excitation source. Fig. 3-20 shows the absorption

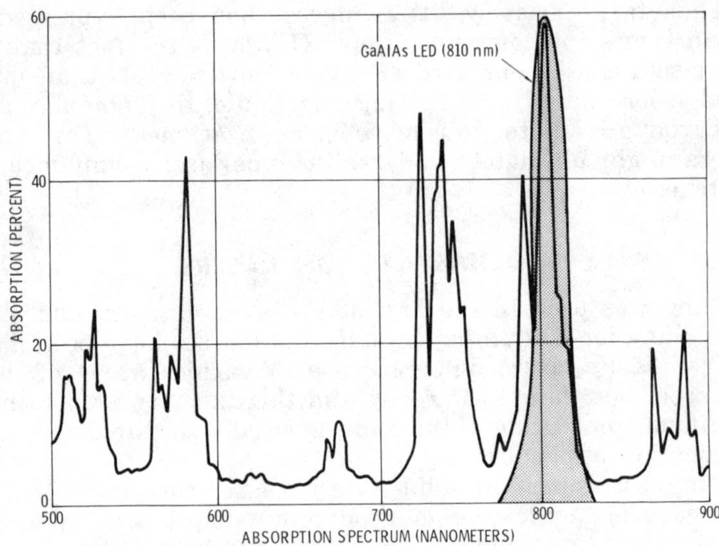

Fig. 3-20. Nd:YAG absorption spectrum versus GaAlAs LED output.

spectrum for a typical Nd:YAG laser rod and also shows the emission from a typical gallium aluminum arsenide LED.

Early diode-pumped YAG lasers consisted of a YAG rod surrounded by rows of LEDs. Recently, however, David A. Draegert of Bell Telephone Laboratories has operated a YAG laser excited by a *single* LED. The construction of this unique laser, which is ideally suited for fiber-optic links, is shown in Fig. 3-21.

The LED has a hemispherical surface designed to emit more radiation than a diode with a flat surface. A gold coating inside a short section of hollow quartz tubing directed approximately half the light from the LED into the laser. The light was kept inside the rod by means of a thin gold layer on the surface of the rod.

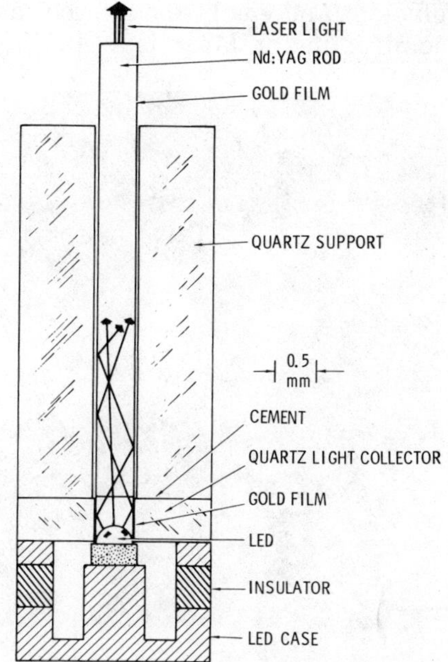

Fig. 3-21. Miniature LED-pumped laser.

The laser rod measured only 0.48 mm in diameter and 5.66 mm long, about the size of a ¼-inch section of a paper clip. Mirror coatings were applied directly to the rod ends. The coating at the diode end was unique in that it was highly reflective at the 1.06-micron output of the laser and was highly transmissive at the 0.81-micron wavelength of the LED.

The single-diode-pumped YAG laser produced only about 100 microwatts when the LED was emitting some 500 milliwatts. While this is a relatively small amount of optical power, improved LEDs and better techniques for coupling the LED radiation into the laser rod will increase this figure.

An even more recent development is a tiny neodymium-doped glass laser pumped by a single injection laser. The laser consists of a glass fiber 1-cm long placed adjacent to the emitting surface of an aluminum-doped gallium-arsenide laser. The first such laser was operated continuously at room temperature, which indicates very high modulation rates are possible. This important new source for optical communications was developed by J. Stone and C. A. Burrus of Bell Laboratories.

SEMICONDUCTOR LASERS

Semiconductor lasers are the smallest and among the most efficient and easily modulated lasers. The tiny size of a typical semiconductor laser is well illustrated in Fig. 3-22, a photo-

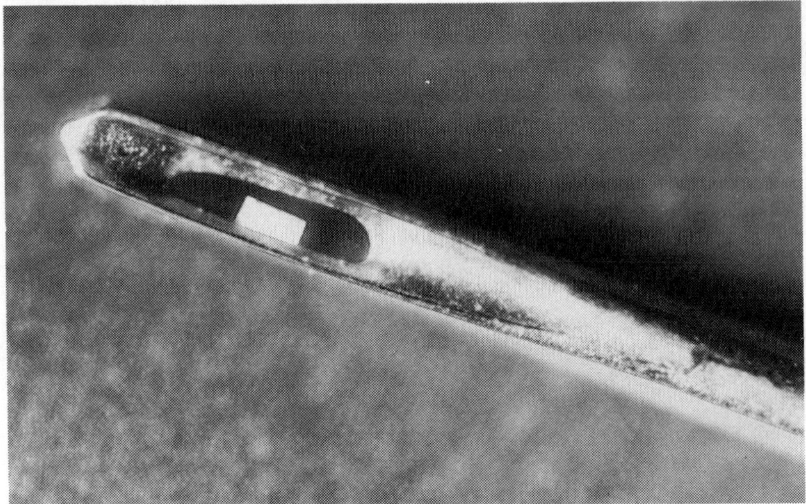

Fig. 3-22. Diode laser inside the eye of a needle.

graph of a gallium-arsenide laser inside the eye of a needle. No less than 38 semiconductors have been operated as lasers.

Various methods for stimulating semiconductor crystals to the lasing state have been employed. The most common excitation methods are avalanche breakdown, electron-beam pumping, optical pumping, and current injection. Of these various methods, direct current injection into a pn-junction diode is by far the most practical for optical-communication applications.

A simple injection laser is shown in Fig. 3-23. The laser consists of a semiconductor chip containing a pn junction formed by diffusion or epitaxy. Two of the opposing ends of the chip are made flat and parallel to each other in order to

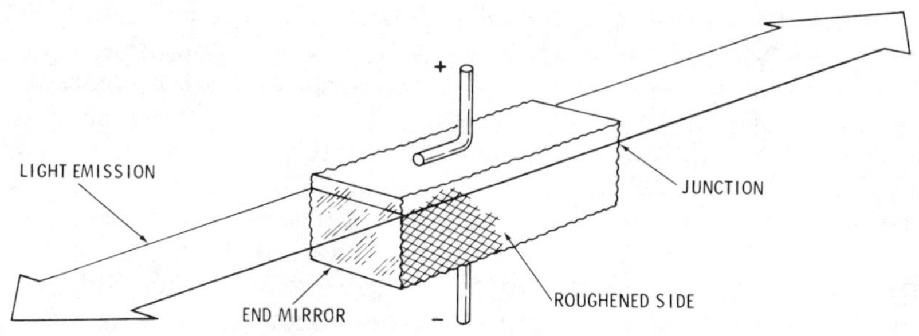

Fig. 3-23. Injection-laser construction.

provide an optically resonant cavity. The other two sides are roughened to suppress unwanted modes. Either the p or the n side of the laser is soldered to a copper heat sink. An electrode wire is then bonded to the opposite side. The laser is packaged in a protective container with a glass window, such as the one shown in Fig. 3-24.

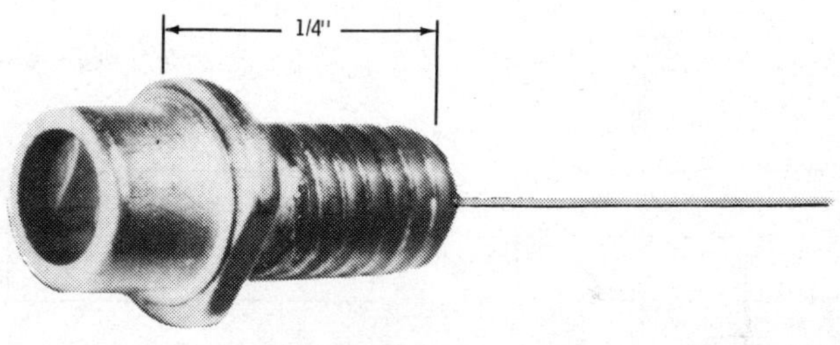

Courtesy RCA

Fig. 3-24. Injection laser mounted in a coaxial package.

Operation of a junction laser is shown in Fig. 3-25. As shown in Fig. 3-25A, an injection laser operated below the lasing threshold acts like an LED and emits light randomly. As more current is injected into the chip (Fig. 3-25B), a point is suddenly reached where a population inversion exists and more atoms are in an excited than unexcited state. Stimulated emission occurs in Fig. 3-25C as spontaneously emitted photons collide with excited electrons in the active region along the pn junction and trigger the emission of additional photons. Finally, in Fig. 3-25D a standing wave of photons is established between the two end mirrors composing the optically resonant cavity necessary for laser action.

The end mirrors of most injection lasers are an integral part of the device. Gallium arsenide (GaAs) and its alloys are the most common injection-laser semiconductors, and their high index of refraction provides a reflectance of approximately 35 percent at the crystal-air interface. GaAs and some other semiconductor laser crystals can be easily given end mirrors by simply cleaving chips of the material along parallel crystalline-cleavage planes to produce very flat and parallel mirror surfaces. Since laser emission occurs from both ends of a chip, a film of gold is often applied to one mirror surface to act as a nearly 100-percent reflector. This causes all the laser light to be emitted from one end of the chip and simplifies external optical requirements. A thin layer of silicon dioxide insulates the gold film from the chip and prevents the pn junction from being shorted.

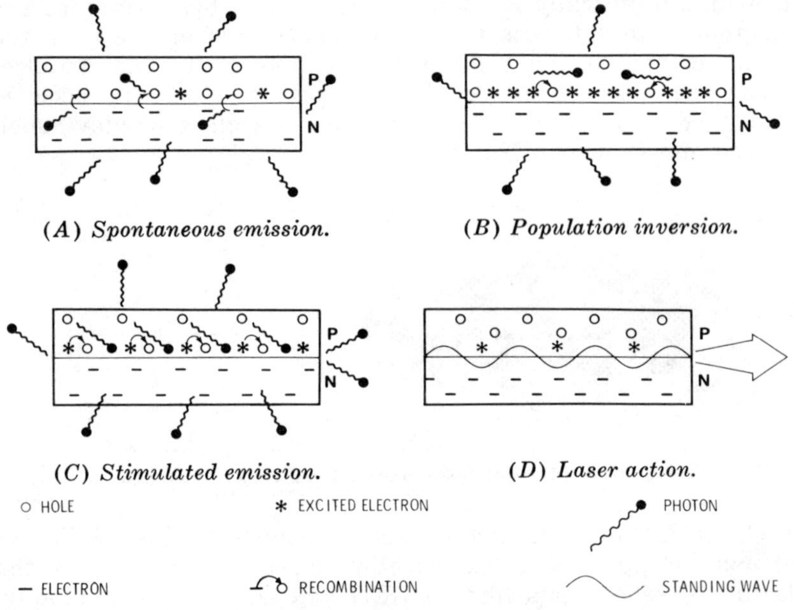

Fig. 3-25. Injection-laser operation.

Depending on their band gap, injection lasers emit at wavelengths ranging from under 700 nm in the visible red to to more than 30 microns in the middle infrared. Gallium arsenide and gallium arsenide alloyed with aluminum are the most common laser semiconductors, and their wavelengths range from 700 nm to 910 nm. The wavelength of all semiconductor lasers decreases with temperature decreases. For gallium arsenide, the change is about 0.25 nm/°C.

The basic junction laser shown in Fig. 3-23 can emit 50 or more watts when pulsed with brief (e.g., 200-nanosecond) current pulses at room temperature. The current density in the junction plane may exceed 100,000 amperes/cm^2; therefore, a typical laser may require a current input of 50 amperes, or more, per pulse.

The very high current densities required by early lasers resulted in rapid degradation of the laser's power output. Furthermore, high-current drivers capable of operation at the high repetition rate necessary for useful optical communications are difficult to design. Therefore, considerable research has been performed to improve the lifetime and reduce the operating current of injection lasers.

Numerous improvements in semiconductor laser technology have resulted from this intensive research effort. One of the most significant developments was the achievement of continuous operation of a diode laser at room temperature by Bell Laboratories scientists in 1970. Prior to this development, injection lasers could be operated continuously only when cooled

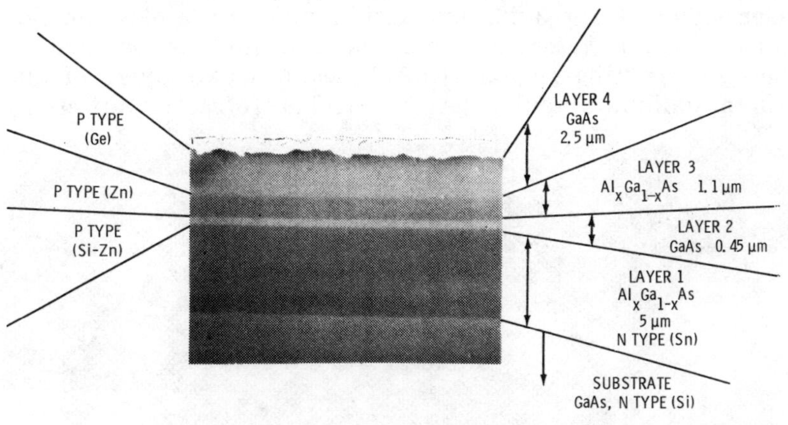

Courtesy Bell Laboratories

Fig. 3-26. A microphotograph of a double-heterojunction injection laser showing the various layers.

with liquid nitrogen. The degradation problem is well understood, and as this is written, Bell Laboratories has operated a diode laser *continuously* and at room temperature for more than a year. The new generation of injection lasers makes practical optical communicators a reality. Indeed, several commercial injection-laser communication systems are described in Chapter 6.

Fig. 3-26 is a microphotograph showing one end mirror of the new Bell Laboratories laser. Unlike the simple pn-junction

Fig. 3-27. Microphotograph of an injection laser capable of operating continuously at room temperature.

laser shown in Fig. 3-23, this laser employs a complex junction called a *double-heterojunction*. The new junction sandwiches the light-emitting region (GaAs) between two layers of aluminum gallium arsenide (AlGaAs). The refractive difference

Fig. 3-28. RCA high-modulation rate large optical-cavity laser diode (center) mounted in a microwave package.

82

between the two materials confines the light to the active region and prevents its escape and absorption into the surrounding crystal. This, coupled with better electron confinement, results in a dramatically lowered operating threshold.

Fig. 3-27 is a photograph of one of the new Bell Laboratories lasers next to a grain of salt. The laser is the small rectangular slab atop the blocklike structure. The larger structure is a tin-plated diamond which performs the role of a high-quality heat sink. Double-heterostructure (DH) lasers have also operated continuously when mounted on very pure copper heat sinks.

Fig. 3-28 shows a DH laser made by RCA mounted in a special microwave package. While this laser cannot be operated continuously at room temperature, its very high repetition rate permits multimegabit communication rates.

A detailed review of the operation of the various kinds of injection lasers is given in *Semiconductor Diode Lasers* by Ralph W. Campbell and Forrest M. Mims (Howard W. Sams & Co., Inc., 1970). This book also gives construction plans for several injection-laser communicators.

CHAPTER 4

Light Detectors

Very few light detectors were available for optical communications before World War II. Selenium was used in most light-beam receivers until the invention of the thallofide cell by T. W. Case in 1917. During World War II, considerable research was conducted by both sides with both solid-state detectors and phototubes.

Thanks to rapid advances in light-detection technology, an almost bewildering variety of optical detectors is available today. This chapter will describe the most important detectors and compare their relative advantages and disadvantages. Those detectors that are being used or considered for use in practical optical-communication links will receive the most attention.

In addition to conventional light detectors, this chapter will also describe several image converters which transform the invisible radiation from an ultraviolet or infrared beam into a visible image. Since many experimental, operational, and proposed optical communicators utilize infrared radiation, this class of detectors is particularly important for beam alignment, pointing, and other operations in which it is convenient to see the transmitted beam.

THE PHOTOELECTRIC EFFECT

The *photoelectric effect* was discovered by Heinrich Hertz in 1887. It is the emission of electrons from a substance illuminated by optical radiation, gamma rays, or X-rays. The effect can be exhibited in several manners. *Photoemission* is the liberation of electrons from a sensitive surface. *Photovoltage* is the potential difference created between two terminals of certain light-sensitive cells usually made from semiconductors. *Photoconduction* is the variation in electrical conductivity in certain materials, usually semiconductors, when they are illuminated by light. Each of these photoelectric effects has found application in light-beam communications.

THE PHOTOTUBE

The phototube is an evacuated or gas-filled tube containing a light-sensitive photocathode and an electron-collecting anode. In operation, light striking the photocathode liberates electrons which are collected by the anode. The result is a *photocurrent* which can be easily amplified. The sensitive photocathode usually consists of ultrathin layers of cesium and cesium-oxide on a cylindrical section of silver. Fig. 4-1 is a diagram of a phototube.

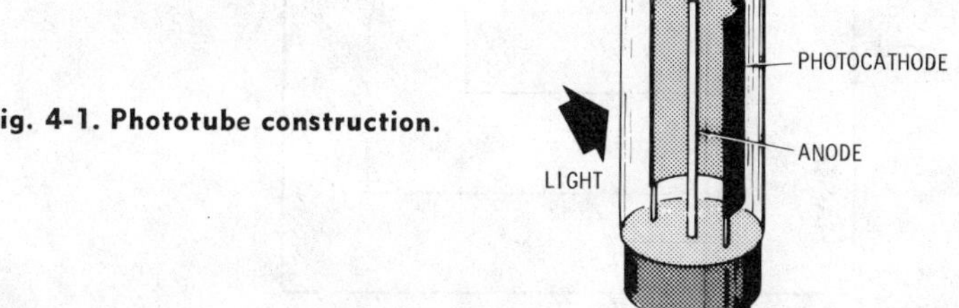

Fig. 4-1. Phototube construction.

Although phototubes have been extensively used in motion-picture sound reproduction and experimental light-beam communicators, they have been surpassed by solid-state detectors and the highly sensitive electron-multiplier phototube.

ELECTRON-MULTIPLIER PHOTOTUBE

The photocurrent of a conventional phototube is quite small, and electronic amplification is necessary for practical applications. The electron-multiplier phototube, or photomultiplier tube, accomplishes amplification of the current flow internally. The photomultiplier tube contains a series of electrons called *dynodes*, each at a higher-voltage potential than the preceding one. Light striking a photocathode liberates electrons which then strike the first dynode. In a process called secondary emission, the dynode emits additional electrons in a ratio of about five electrons for every incoming electron. The secondary electrons continue on to the next dynode where still more electrons are liberated, and the process continues until the second-

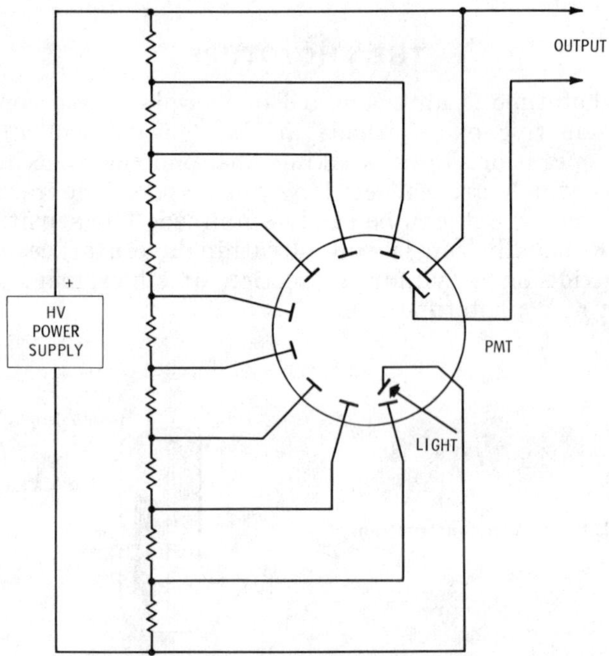

Fig. 4-2. Photomultiplier operating circuit.

ary electrons strike the anode and produce a photocurrent very large in proportion to the light striking the photocathode. Since as many as ten or more dynodes are employed in a photomultiplier tube, the electron gain is considerable. Fig. 4-2 is a diagram of the electrical connections made to a typical photomultiplier tube. The resistor chain acts as a voltage divider which supplies the sequential voltage levels to each successive dynode.

Photomultipliers are the most sensitive detectors available for most of the optical spectrum. They are available in a wide range of spectral sensitivity ranges and internal amplification figures, and they have exceedingly fast rise times. Photomultiplier response can be so good that individual photons can be detected and counted.

Although photomultipliers rank among the most sensitive optical detectors available, they are disadvantaged by their requirement for a high bias voltage. Additionally, their large size and fragile construction are serious drawbacks in the construction of compact, sturdy optical communicators. Finally, the high sensitivity of photomultipliers means they are very susceptible to interference from unwanted light sources, and special narrow-band optical filters must be employed when the photomultiplier is used during daylight or in the presence of an artificial light source.

For these reasons, photomultipliers are usually restricted to very-long-range space and atmospheric communicators. Solid-state detectors, such as the avalanche photodetector, are used in most short- to medium-range applications.

PHOTORESISTORS

Numerous types of inexpensive photoresistors are available. Like Bell's selenium cells and Case's thallofide cells, all of these devices are characterized by a resistance that is high in darkness and low in the presence of light.

Photoresistors can be very sensitive, with dark to light ratios of more than 10,000:1 being common. But their very slow response time severely limits their usefulness in practical light-beam communications. The most common photoresistors for detection of visible light utilize thin layers of cadmium

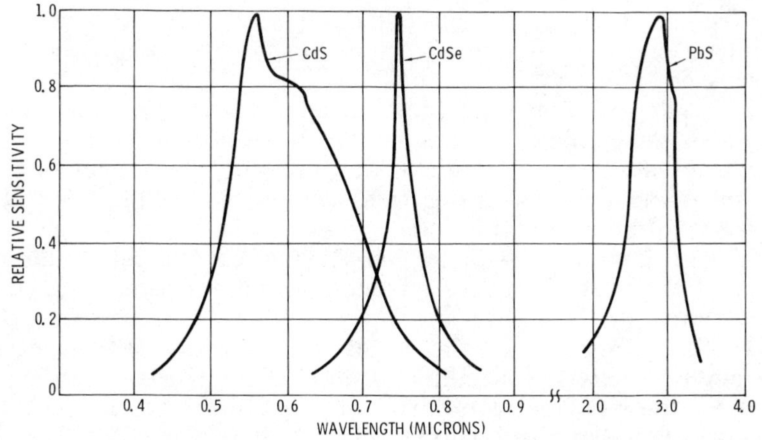

Fig. 4-3. Spectral response of several common photoresistors.

sulfide (CdS) or cadmium selenide (CdSe) on an insulating substrate. Electrical connections are made to the material, and a protective package with a transparent window is provided. Fig. 4-3 shows the spectral response of typical CdS and CdSe compositions. Fig. 4-3 also shows the response of lead sulfide (PbS), a photoresistor well suited for the detection of near-infrared radiation.

Most photoresistive light detectors have a very slow response time, and some are affected by light-history effects. In the latter phenomenon, a light level of sufficient intensity temporarily alters the sensitivity of the cell so that its response to a subsequent light level is impaired. Light-history effects may last seconds or even minutes.

SEMICONDUCTOR PHOTODIODES

Variations of the pn-junction diode comprise the most important family of detectors available to the optical-communications engineer. The family includes photovoltaic detectors, such as solar cells; photoconductive detectors, such as reverse-biased pin detectors; and avalanche detectors. First, the operation of photodiodes will be reviewed, and then each of these detector types will be discussed separately.

All semiconductor junction devices possess some degree of light sensitivity. In fact, plastic-encased transistors of the early 1950s were found to be sensitive to external light and it was necessary to add light-absorbing material to the plastic. Common glass-encased diodes are to a certain extent light sensitive, and even light emitting diodes are light sensitive. Since LEDs can both emit and detect optical radiation, several unique communication applications are possible, and these are discussed later in this chapter.

Light detection at a pn junction occurs when photons striking the semiconductor create hole-electron pairs. The pairs can be created on either side of the junction. The energy of the illuminating photons causes the electrons to be displaced upward into the conduction band, while the holes remain behind in the ground state (valence band). The process is similar to that of the excitation of electrons at a pn junction by a current flow.

If the two sides of the junction are connected together through an external circuit or load, the excited electrons will resume their normal equilibrium by combining with holes on the other side of the junction. The result is a current flow as both electrons and holes move toward the junction.

88

The process just described occurs in a photovoltaic detector. The result is a current flow without the assistance of anything other than light. This phenomenon is utilized in the silicon solar cell to generate electricity for earth satellites, automated weather stations, and other specialized systems. As the cost of the solar cell is lowered, more applications will be found for this important photovoltaic detector.

Photovoltaic detectors can be used in simple light-beam communication receivers. It is possible, for example, to connect a silicon solar cell directly to a transistor amplifier to make a simple but effective light-beam receiver. Fig. 3-1 shows a receiver made in this manner.

Photovoltaic detectors can be made more effective, however, by biasing the diode and permitting it to act as a current source. If the diode is connected in the normal or forward direction, the current increase will not be great. But if the junction is connected in the backward direction, there will be a substantial flow when light is allowed to illuminate it. This connection method is called *reverse biasing,* and it is frequently employed with silicon photodiodes. Since the photodiode reverse resistance *decreases* (current flow increases) with applied light, a reverse-biased photovoltaic detector is said to be operated in the *photoconductive* mode. Fig. 4-4 shows a photo-

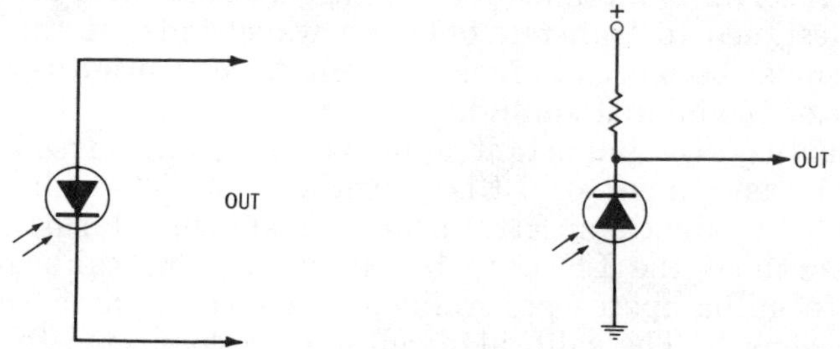

Fig. 4-4. Photovoltaic and photoconductive photodiode operation.

diode connected in both the forward and the reverse direction. The load resistor provides a means for extracting the photocurrent from the circuit. The voltage across a large-value load resistor is higher than that across a small-value resistor, so large values of R_L are used for high sensitivity. Since a large value for R_L slows the response time for a detector, sensitivity must be traded off against frequency response.

A photovoltaic detector operated in the photoconductive mode resembles the light-induced variable resistance of the photoresistor. However, under constant illumination the photoconductor acts as a constant-current source, no matter what the bias voltage is. The current in a photoresistor varies with the incident-light level. Summing up, reverse biasing provides twice the responsivity of photovoltaic operation and far greater response time.

SOLAR CELLS

Solar cells are large-area (up to several square centimeters) diffused silicon pn diodes capable of generating a photocurrent of up to 100 milliamperes at a potential of 0.5 volt when illuminated by bright sunlight. Because of their low cost and their availability, conventional silicon solar cells are well suited for use in short-range communication systems, such as the one shown in Fig. 3-1.

Most silicon solar cells have a peak spectral response of about 850 nm in the near-infrared region. This closely approximates the peak-emission wavelength of aluminum-doped gallium-arsenide emitters. The rise time of a typical 2-×2-cm cell is about 20 microseconds, more than adequate for detection of audio-modulated light beams. The large active area of most silicon cells makes them useful in short-range optical communicators which employ no receiver optics. However, unless such a receiver is intended for operation in the dark, an optical filter designed to transmit only the wavelength of the transmitter must be placed before the cell. An exception is the detection of modulated sunlight.

A particularly important application of silicon solar cells is the measurement of LED power output. The cell is first calibrated against a spectral source of known intensity at the wavelength of the LEDs to be measured. The calibration is given in milliamperes per milliwatt and varies as a function of wavelength. The calibrated cell is then connected to a very low resistance load (e.g., 0.1 ohm), and a microvoltmeter is used to measure the voltage across the load. The cell current is then calculated according to Ohm's law, and the LED power is calculated from the cell calibration factor.

PIN PHOTODIODES

Pin diodes are finding widespread use in optical communicators employing LEDs and injection lasers. The two most common pin-diode configurations are the Schottky-barrier and planar-diffused types. Both structures consist of heavily doped p and n regions separated by an undoped region of bulk (intrinsic) silicon. The resistivity of the bulk layer can range from 0.1 ohm-cm to 100,000 ohm-cm.

When an external bias is connected to a pin photodiode in the reverse direction, an electric field is established within the intrinsic region. With no light present, a *dark current* of approximately a microamp will flow. The dark current is undesirable since it determines the minimum detectable signal level, and better cells having very low dark currents are available. As external illumination falls on the sensitive surface of the cell, the diode passes a much larger photocurrent. A typical pin photodiode will pass a current of from 0.4 to 0.5 microampere per microwatt of applied illumination at the peak spectral response of the diode (usually about 900 nanometers). This figure of merit is called *responsivity*. Some diodes are made even more sensitive by the addition of a thin antireflection coating to the sensitive surface. The coating permits some radiation that normally would be reflected at the silicon-air interface to be absorbed in the cell.

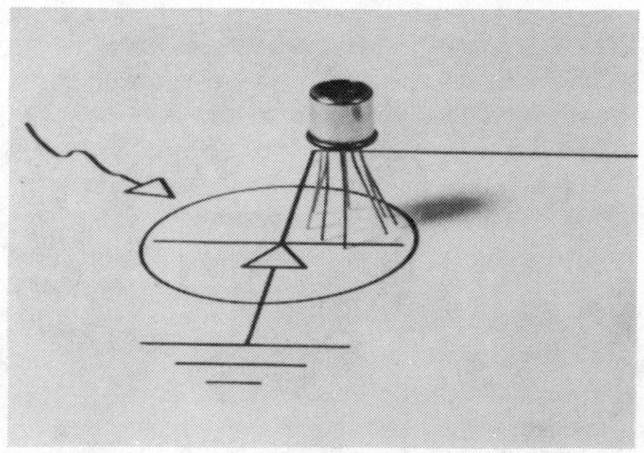

Courtesy United Detector Technology

Fig. 4-5. Typical pin diode-amplifier module.

Pin diodes can be fabricated in a wide variety of configurations. Mosaics, line configurations, and quadrant detectors are commonly employed in such applications as counting, sensing, surveying, and many others. Most optical communicators employ single detectors having a circular format. Several manufacturers offer pin diodes installed in a small transistor or integrated-circuit-type package complete with amplifier circuit. These detector modules can greatly simplify the design and construction of an optical receiver. Fig. 4-5 is a photograph of a detector/amplifier module.

Pin-diode modules incorporating an operational amplifier are characterized by an output that is linear over at least seven decades of incident-light intensity. This permits a pin-diode/operational-amplifier module to be employed as a linear optical power meter by simply connecting a meter to the output and a series of calibrated resistors to a feedback-resistor range switch.

Pin photodiode modules are manufactured by RCA; EG&G; MERET, Inc.; United Detector Technology; Bell and Howell; and other firms. One of these devices, the MERET FDA425, measures only 9.4 mm square and has a sensitivity of 70 nanowatts with a snr of 10:1 at the 905-nanometer wavelength from a GaAs LED.

SCHOTTKY-BARRIER PIN PHOTODIODES

A large-area Schottky-barrier pin photodiode is shown in Fig. 4-6. The device is made by applying an aluminum layer

Fig. 4-6. Schottky-barrier pin photodiode.

to one side of a slab of intrinsic silicon and a transparent gold film to the other side. Both metal layers serve as contact points, but the parameters of the gold layer are critical since the light to be detected must pass through it. The gold layer is usually an evaporated film only about 150 angstroms thick. The thickness of the silicon wafer can vary from about 0.125 to 1.0 millimeter.

For the Schottky-barrier diode to operate in an optimum manner, an electric charge should extend across the intrinsic bulk silicon to both p and n regions at the contacts. When this occurs, the diode is said to be fully depleted. The thickness of the intrinsic region is directly related to the voltage required for depletion. To calculate the penetration (depletion depth) of the field, a simple formula can be used:

$$d = \frac{\sqrt{RE}}{2}$$ (Eq. 4-1)

where,

d is the depletion depth in microns,
R is the silicon's resistivity,
E is the applied reverse-bias voltage.

This formula can be valuable in calculating required voltage for full depletion when both resistivity and the intrinsic-region thickness are known.

DIFFUSED PIN PHOTODIODES

The structure of a diffused pin photodiode is probably familiar to most readers. The back contact side of both kinds of diodes consists of an evaporated aluminum layer. However, the diffused diode incorporates conventional masking and diffusing techniques for the front side. First, a layer of silicon dioxide is evaporated over the intrinsic silicon wafer. Then, a hole is etched to reveal some of the bulk silicon. Next, an impurity such as boron is diffused into the bulk material to form a junction. Contact is made directly to the high-conductivity boron diffused region by means of one or more bonded wire electrodes. Fig. 4-7 shows a large-area diffused pin diode.

GUARD RINGS

Both Schottky-barrier and diffused pin photodiodes exhibit a dark current, and since the ultimate sensitivity of a photo-

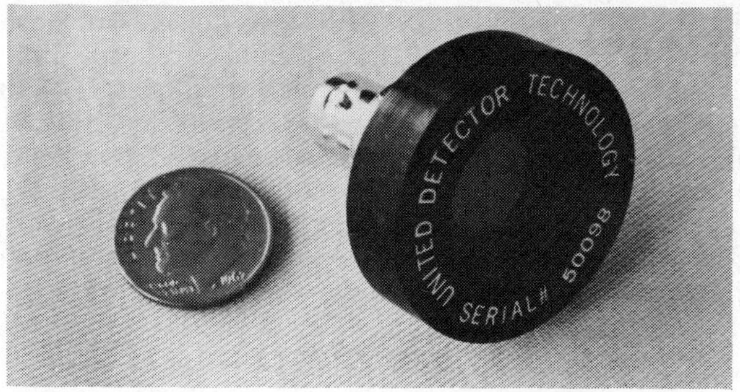

Courtesy United Detector Technology

Fig. 4-7. A large area diffused pin photodiode.

diode is dependent on a high signal-to-noise (snr) ratio, it is important to reduce current leakage to an absolute minimum. Part of the dark current is due to bulk leakage, but some is caused by surface leakage over the sides of the diode and across the junction. Surface leakage, sometimes called channel leakage, can be eliminated by means of special fabrication techniques or with an additional electrode called the guard ring. The guard ring is fabricated at the same time as the active-area junction. On a diffused pin photodiode, a circle is etched in the silicon-dioxide layer around the active region and both are diffused simultaneously. In fact, both regions become pn junctions. The guard ring is activated by biasing it at the same voltage as the active area. This causes any surface leakage to flow through the guard ring and not the active area. Since bulk leakage is as much as 100 times less than the surface leakage, the guard ring can significantly improve the performance of a pin diode by increasing its snr.

SCHOTTKY-BARRIER VERSUS DIFFUSED PHOTODIODES

As might be expected, both Schottky-barrier and diffused pin photodiodes have relative merits and demerits. Schottky-barrier diodes can be constructed in much larger sizes than their diffused counterparts, but they are more temperature sensitive and have less immunity to high light levels than diffused diodes do. The difference in maximum acceptable light-intensity levels is due to the possibility of damage to the gold-film electrode over the active area of Schottky-barrier devices. On the other side, the heat required to form diffused diodes can alter the characteristics of the high-resistivity silicon intrinsic region.

Both types of diodes have very low noise, and both have response times measured in nanoseconds. A very important characteristic possessed by these types of diodes is their high degree of linearity, and both Schottky-barrier and diffused diodes are linear over about seven decades of light intensity. This feature is particularly important in applications where light levels will vary greatly. It is also very important in optical receiver systems designed to be immune from the effects of sunlight without the need for an external filter.

AVALANCHE PHOTODIODES

If the reverse bias on a diffused pin photodiode having a very thin intrinsic region is increased beyond a certain point, the diode may break down and permit a current to flow. If the bias is maintained just below the breakdown region, the diode becomes very sensitive to light. Incoming photons create a disproportionate number of hole-electron pairs, and the result is called an avalanche multiplication process. Fig. 4-8 is a photograph of an RCA avalanche detector.

Fig. 4-8. RCA silicon avalanche detector.

Internal gain is the most important characteristic of the avalanche detector. A conventional pin photodiode has a sensitivity of about 0.5 milliampere per milliwatt, while an avalanche detector may have a sensitivity of 100 or more milliamperes per milliwatt.

Since the avalanche diode has internal gain, it is capable of detecting very low light levels. Commercially available detectors, for example, can detect an optical signal having an amplitude of a nanowatt or less. This high degree of sensitivity is why the avalanche detector is the only optical detector that rivals the photomultiplier.

For all its impressive abilities, the avalanche detector has a big drawback—the operating voltage must be carefully maintained just below the detector's breakdown point. Since the avalanche point varies with temperature, a temperature-compensating bias supply must be employed for best results. The design of such a supply can be difficult, so several manufacturers market avalanche photodetector modules that contain an integral temperature-tracking power supply and a preamplifier. These modules are manufactured by American Laser Systems, Inc.; General Electric; Texas Instruments; and other firms. Some modules include a thermoelectric cooler to improve detector sensitivity.

A typical avalanche-photodetector module is the American Laser Systems Model 728. This system consists of a compact detector module and preamplifier measuring only 2.9 cm × 3 cm. A separate temperature-tracking bias supply provides a regulated high-voltage bias for the detector from a 12-volt battery. The system has a sensitivity of 0.58 nanowatt at 900 nanometers. The American Laser System avalanche-detector system is employed in several medium-range laser-communication systems, two of which are described in Chapter 6.

Avalanche-detector systems are at least ten times more costly than pin-diode detectors. Typical pin diodes cost ten to twelve dollars, while avalanche-detector modules sell for a few hundred dollars. Of course, the avalanche diode itself is available at less cost than the module, but the difficulty of fabricating a suitable temperature-tracking bias supply makes it worthwhile to consider purchasing the entire module.

Their higher cost notwithstanding, avalanche detectors offer an excellent detector for operational optical-communication links employing LEDs, injection lasers, or Nd-doped YAG lasers. One manufacturer reports a trend toward communicators employing LEDs and avalanche detectors instead of injection lasers and conventional pin diodes. The avalanche-detector system offers such high sensitivity that the higher output of the laser is just not needed in some moderate-range communication links. The higher initial cost of the avalanche detector is offset by the increased reliability of the LED over the injection laser.

PHOTOTRANSISTORS

A great variety of both bipolar (usually npn) and field-effect phototransistors is available. These devices employ a modified construction to permit light to strike the sensitive regions. Most early phototransistors were housed in conventional metal transistor cans with a flat glass window or lens in one end, and this type of packaging is preferred for applications where mechanical strength is important. More recently, all-plastic packaging has become popular, and plastic units are available from many manufacturers at low cost.

Fig. 4-9 shows several miniature phototransistors manufactured by Sensor Technology.

The operation of a bipolar phototransistor can be considered as similar to that of a photodiode having internal gain. If the base-collector junction is exposed as the light-sensitive region, any light-induced current will become the base current of the transistor. Since no external base-bias current is re-

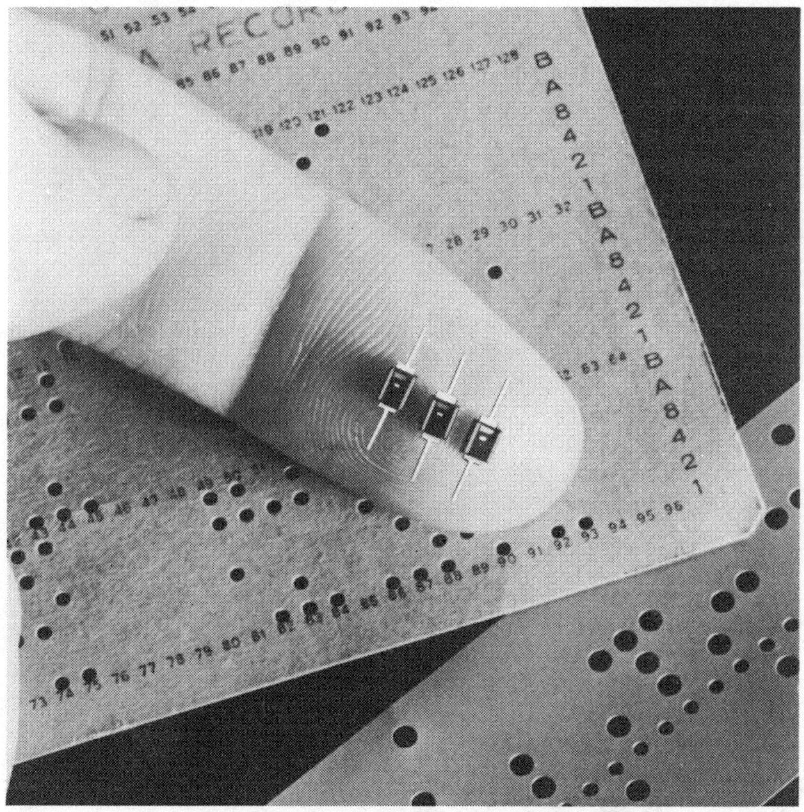

Courtesy Sensor Technology, Inc.

Fig. 4-9. Miniature encapsulated phototransistors.

quired for amplification to occur, many phototransistors are made without a base lead. Base bias is supplied by the incoming optical signal. Some applications, particularly those requiring high sensitivity, benefit from the use of an external base electrode since this permits conventional transistor biasing to be used. A prebiased phototransistor is more sensitive to incoming radiation than an unbiased unit. Phototransistors

97

without a base lead can be biased by illuminating them with a low-level, dc light source.

The chief advantage of a phototransistor over a photodiode is gain, and phototransistors may have a sensitivity of a few milliamperes per milliwatt. Unlike the photodiode, however, the phototransistor does not have a linear gain and this results in rapid saturation. For very small incoming-light levels, the gain is reasonably linear, but saturation rapidly occurs as the incident illumination passes a certain intensity. As a result, phototransistors cannot be used to detect modulated light beams in the presence of high ambient-light levels, unless an optical filter is used to eliminate a substantial percentage of radiation from the interfering source.

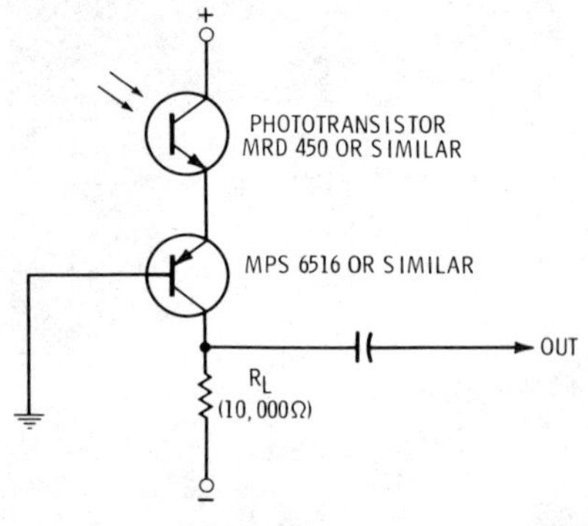

Fig. 4-10. Phototransistor speed-up circuit.

Slow response time is another disadvantage of the phototransistor, although megahertz operation can be obtained by employing special circuitry to reduce the effect of the device's capacitance.

Fig. 4-10 shows one way to improve the response time of a phototransistor. This circuit uses a large load resistor to give high sensitivity, yet it preserves the low impedance necessary to give the phototransistor a fast response time. The parts values shown are suggested values, and substitutions are permissible. Maximum frequency response will occur when the load resistor has a value of 10,000 ohms. More sensitivity can be obtained by increasing this resistance, but an increase beyond about 1,000,000 ohms will cause a significant reduction in frequency response.

LIGHT-SENSITIVE FETs

The light-sensitive field-effect transistor has been used in several experimental injection-laser communication receivers.

Teledyne Crystalonics manufactures light-sensitive FETs under the trade name FOTOFET. The FOTOFET has a very high impedance, very fast response time, and high sensitivity. A typical FOTOFET will exhibit a response time of 30 nanoseconds—approximately midway between that of phototransistors and pin photodiodes.

LEDs AS DETECTORS

An LED can be used to detect the recombination radiation of an LED having a similar band-gap energy. This requirement is easily solved by using a detector LED fabricated from the same semiconductor as the emitter LED.

The significance of an LED's detector ability is particularly important in optical communications. For example, the author has constructed an optical communicator which permits two-way simplex communications with a system employing a single LED and lens at each end of the link. The LED is switched to the output of an amplifier in the transmit mode and to the input of the same amplifier in the receive mode. Since the field of view of the LED in the receiver mode is identical to, and coaxial with, the field of illumination of the LED in the transmit mode, optical alignment of the system is considerably simpler than that of communicators employing separate transmitter and receiver optics.

The author has measured the quantum efficiency of an RCA SG1004 infrared-emitting GaAs:Si diode by illuminating the diode with a second SG1004. The quantum efficiency of the detector LED was slightly higher than 50 percent, and even better results can be expected with better coupling between the two LEDs. Silicon pin diodes may have quantum efficiencies of 80 percent or more—but they do not have the ability to both detect and emit radiation.

The LED–LED optical communicator just described provides a practical demonstration of LED sensitivity in the detector mode. This amplitude-modulated system can detect the optical signal from an identical LED when only about 200 nanowatts strike the lens of the detector LED.

The injection laser, which is actually a specially configured LED, can also be employed as an optical detector. The author has used an injection laser to detect audio signals from a similar laser. The experiment was performed in a laboratory environment, but there is no reason why the principle cannot be extended to long-range applications. Optical-fiber coupling is a particularly attractive role for semiconductors that both emit and detect radiation since data can be transmitted two ways over a single fiber.

IMAGE CONVERTERS

Image converters are useful for viewing the radiation from near-infrared optical communicators. Phosphor screens, which convert the invisible near-infrared radiation from GaAs LEDs and injection lasers to a visible orange glow, are available at low cost and are very useful for aligning the optics of a communicator. Viewing devices that utilize image-converter tubes are required to sight a transmitter beam from a distant receiver location.

IMAGE-CONVERTER TUBES

From the previous brief discussion on vacuum phototubes, it will be recalled that photoemissive substances have the property of emitting electrons in response to an oncoming beam of light. If one end of an evacuated tube is coated with photoemissive material and if a high-voltage electrostatic field is applied across the tube, it is possible to direct the electrons emitted when the photoemissive surface is struck by light, back toward the end of the tube. With suitable regulation of the electrostatic field, the electron pattern at the rear of the tube will duplicate the light pattern at the light-sensitive face of the tube. These electrons can be used to stimulate a phosphor-coated screen at the rear of the tube.

If the photoemissive surface is sensitive to infrared and if the viewing phosphor converts the energy of impacting electrons into visible light, the image tube described above becomes an image converter. That is, it converts invisible infrared into visible light. Typical image-converter tubes require 15,000 volts dc at low current for proper operation. Some tubes incorporate a central electrode called the *focusing electrode*. Others achieve self-focusing by means of suitably shaped primary electrodes.

CHAPTER 5

Optical Components and Systems

The range capability of an optical-communication system largely depends on the optical components employed to collimate the transmitter beam and to collect and focus this radiation upon the receiver's detector. The most important optical components are lenses, reflectors, filters, and fiber waveguides. These and other optical components are described in this chapter.

LENSES

A lens is a transparent device that relies on geometry and refraction to manipulate a beam of light. Most lenses are made from glass or plastic, but middle- and far-infrared radiation requires lenses made from germanium or other exotic infrared-transmitting materials.

Glass lenses are generally more costly than their plastic counterparts. Though plastic lenses do not have the very high optical quality of most glass lenses, they are lightweight and shatterproof. Plastic lenses are well suited for use in short-range communicators where superior optical quality is not required. At least one manufacturer has used a plastic lens containing an infrared-transmitting dye in a plastic receiver lens to filter out unwanted external illumination.

Converging (positive) lenses have one or both surfaces curved outward in either a convex or a double-convex configuration. The converging lens collects light and focuses it into a small spot called the focal point, as shown in Fig. 5-1. Converging lenses can also collect light emerging from a point and collimate it into a narrow beam.

Converging lenses are widely used as both transmitter and receiver antennas in optical communicators. Equation 2-1 in Chapter 2 permits the beam divergence of a light source-lens

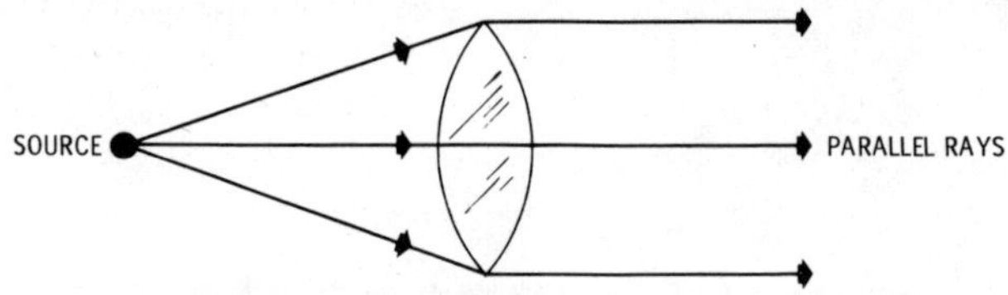

(A) Collimating radiation from a source.

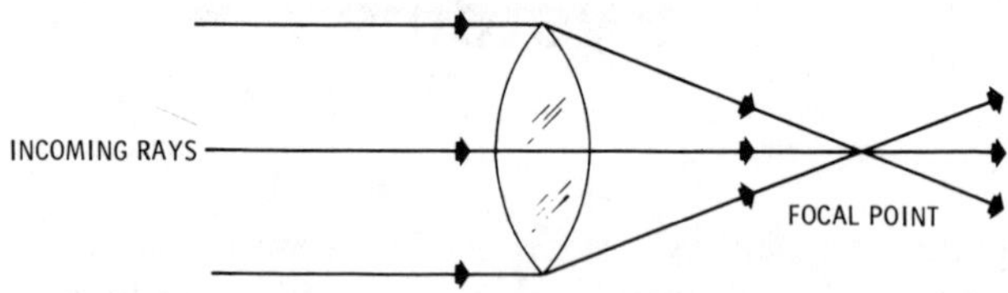

(B) Focusing radiation from a distant source.

Fig. 5-1. Converging (positive) lens ray diagrams.

combination to be calculated when the lens focal length and source diameter are known:

$$\theta = \frac{d}{f} \qquad \text{(Eq. 2-1)}$$

This relationship shows that the divergence of a source-lens pair becomes smaller as the lens focal length is increased or as the diameter is decreased. For example, according to Equation 2-1, an LED having an emitting region 1 mm in diameter will produce a beam having a divergence of 40 milliradians when paired with a lens having a 25-mm focal length. An injection laser with an emitting region measuring 0.5 mm by 0.002 mm will produce a beam having a divergence of only 20 mr by 0.08 mr when paired with the same lens.

When the source size is fixed and when it is desirable to reduce the divergence of the transmitted beam, a lens with a longer focal length can be employed. However, unless the diameter of the lens is also increased, less light will be collected by the lens. In many cases it is desirable to use "fast" lenses having a small f/number (see Equation 2-3).

In using Equation 2-1, one must remember that the actual divergence will be increased somewhat by lens imperfections, atmospheric effects, and diffraction. Diffraction is caused by the finite size of the lens, and part of any light beam passing through an aperture will be diffracted outward and away from the main beam. Diffraction usually takes the form of several concentric rings surrounding the central beam. Usually, the power contained in the diffracted portion of a light beam is not high, and the use of a lens several centimeters or more in diameter will help reduce diffraction effects. Fig. 5-2 illustrates how diffraction occurs at an aperture.

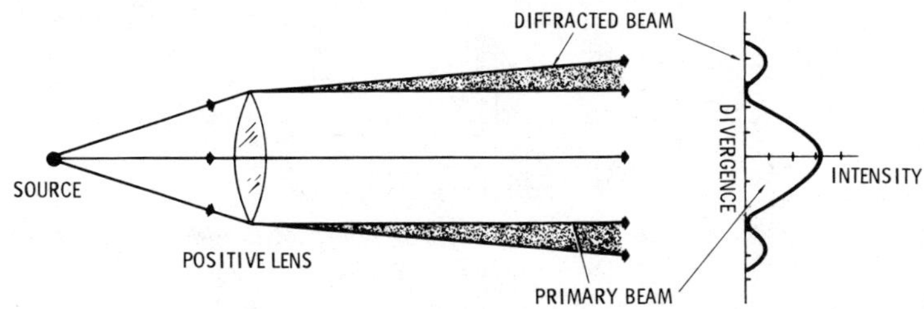

Fig. 5-2. Diffraction of a light beam.

Some photodetectors have a very small active area, and it is necessary to apply another lens formula to determine the spot size of a light beam focused by a converging lens. The spot size is:

$$d_s = \frac{2.44f\tau}{d_R} \qquad \text{(Eq. 5-1)}$$

where,

  d_s is the spot diameter,
  f is the lens focal length,
  τ is the wavelength of the radiation being focused,
  d_R is the diameter of the receiver lens.

Diverging (negative) lenses are also used in some optical communicators. The diverging lens accepts an incoming light beam and spreads it outward as shown in Fig. 5-3. Conversely, a diverging lens can convert a converging beam into a parallel beam. The latter property is very useful when a narrow-band interference filter is employed to block unwanted light at the detector of the receiver. As shown in Fig. 5-4, the diverging lens causes radiation collected by the primary lens to pass through the filter in a parallel beam. This is important because an interference filter will not operate properly if nonparallel light is passed through it.

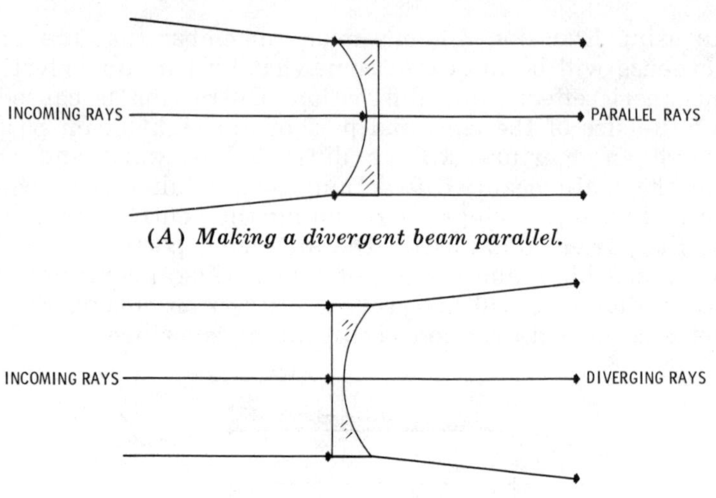

(A) Making a divergent beam parallel.

(B) Diverging a parallel beam.

Fig. 5-3. Diverging (negative) lens ray diagrams.

The cylindrical lens is a special-purpose optical component which bends light rays in only two dimensions so that a linear focal region, rather than a circular focal spot, is formed. As shown in Fig. 5-5, cylindrical lenses are very useful for collimating radiation from linear light sources such as some injection lasers. They are also useful for focusing radiation upon linear detectors and linear-detector arrays.

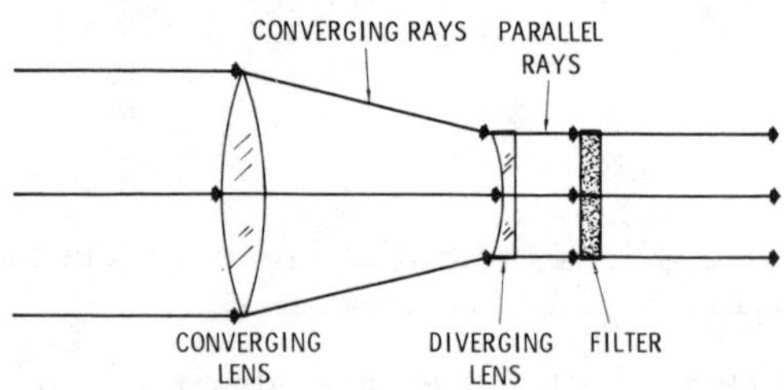

Fig. 5-4. Use of a diverging lens with an interference filter.

Thus far, we have described *simple* lenses. Two or more simple lenses can be cemented together with a transparent cement to form a *compound* lens. Two bonded simple lenses form a *doublet*, and three form a *triplet*. Canada balsam is frequently used to bond lenses.

A particularly important compound lens is the *color-corrected achromat*. As shown in Fig. 5-6, the focal point of a simple converging lens varies slightly with wavelength. There-

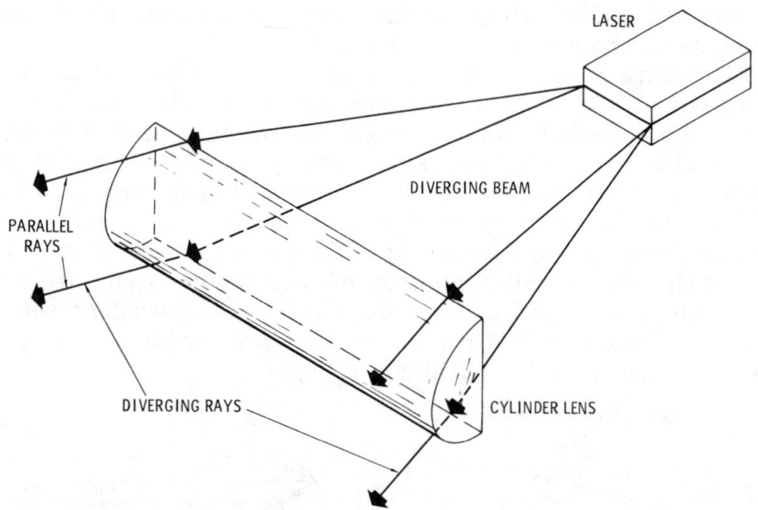

Fig. 5-5. Use of a cylinder lens with a laser diode.

fore, nonmonochromatic light cannot be focused to as small a spot as monochromatic light (see Equation 5-1). Similarly, nonmonochromatic light cannot be as perfectly collimated by a simple lens as monochromatic light can.

The color-corrected achromat employs a compound converging-diverging (positive-negative) lens pair to give a focal length that is independent of wavelength. The effectiveness of the achromat is shown by the beneficial results obtained when it is used in optical communicators employing semiconductor injection lasers. Even though the spectral width of an injection laser may be 4 nanometers or less, an achromat provides measurably better focusing and collimating ability than what a simple lens provides. The effects are even more pronounced with LED systems since their spectral width may be 25, or more, nanometers.

PARABOLIC REFLECTORS

Parabolic reflectors are important transmitter antennas for wide-angle light sources, such as incandescent lamps, glow

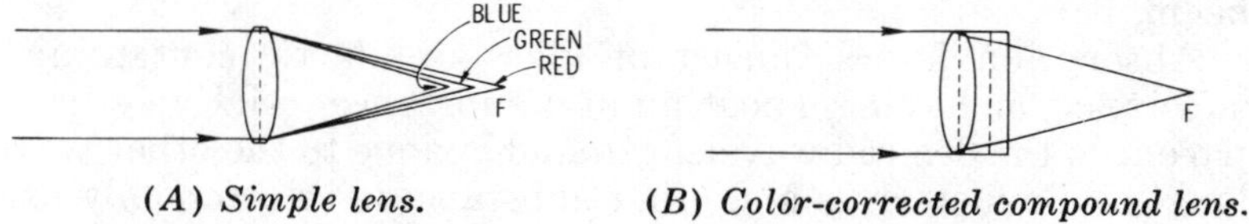

(A) Simple lens.   *(B) Color-corrected compound lens.*

Fig. 5-6. Improved focusing with color-corrected achromat.

tubes, and LEDs. They are also important receiver antennas because of their small f/number.

Miniature parabolic reflectors are an integral part of some high-power LEDs. The reflector can more than double the radiation collected from some LEDs. Because of their smallness, miniature reflectors can be used in conjunction with external lenses to further reduce the beam width from an LED.

The antenna gain of parabolic reflectors can be very high, particularly in the case of transmitter antennas. This is because the very small f/number of a parabolic structure permits far more light to be collected from a divergent source than can be collected with a lens. Fig. 5-7 compares the operation of a lens and a parabolic reflector.

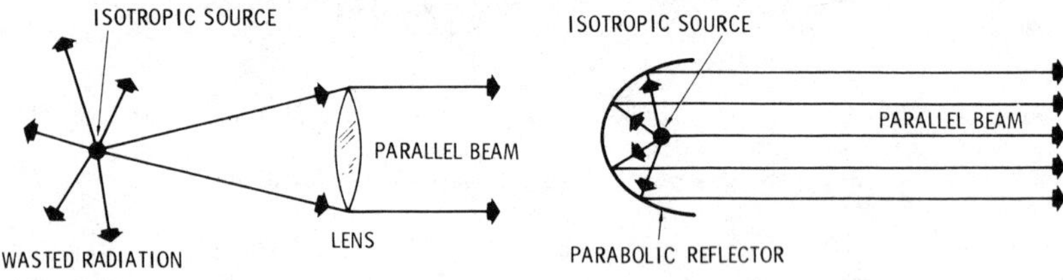

(A) *Low-gain projection system.*   (B) *High-gain projection system.*

Fig. 5-7. Comparison of lens and parabolic reflector projection system.

Most parabolic reflectors are made of polished metal or of plastic or glass coated with a thin film of highly reflective aluminum or gold. Gold is generally used for near-infrared wavelengths because of its superior reflectance. Plastic reflectors are potentially very economical and are finding use as reflectors in low-cost flashlights. They are considerably less fragile and lighter in weight than comparable glass reflectors.

FILTERS

Narrow-bandpass optical filters are frequently used in optical-communication receivers to block unwanted light from the detector. Filters are sometimes used to block the visible light from an incandescent source so that only the infrared wavelengths are transmitted. This provides an invisible, covert beam.

Absorption filters consist of glass or plastic containing a substance, or having a coating of a substance, relatively transparent to the desired wavelength and opaque to the other wavelengths. Various dyes and other substances are commonly used to make absorption filters. Since the dye may absorb as much as half of the desired wavelength, absorption filters are relatively inefficient. Nevertheless, they are reasonably priced and find use in some operational light-beam communicators.

Interference filters have multilayer dielectric coatings to selectively reflect unwanted wavelengths while transmitting the desired wavelength with a relatively high degree of efficiency. The layers consist of very thin films separated by precise thicknesses of inert material. Since the manufacturing process for interference filters is complex, they are generally more costly than absorption filters. However, their filtering efficiency is far superior to that of absorption filters since up to 95 percent of the desired wavelength is transmitted. Additionally, interference filters pass a very narrow spectral range as compared to absorption filters.

Fig. 5-8 shows a miniature infrared interference filter. This filter is designed to transmit the wavelengths emitted by a GaAs injection laser, and its spectral transmission curve is

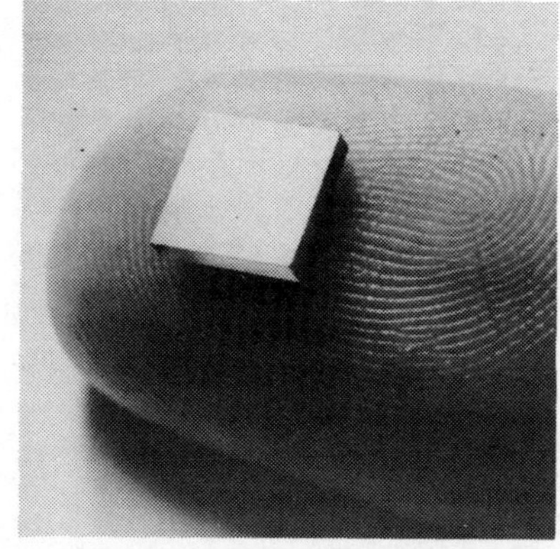

Fig. 5-8. Infrared-interference filter.

Courtesy Optical Coating Laboratories, Inc.

shown in Fig. 5-9. Fig. 5-9 also shows the transmission curve of a typical infrared absorption filter. The advantages of the interference filter over the absorption filter are made obvious by the figure. The major advantage of the absorption filter is its moderate cost.

Some interference filters employ an absorption filter as a substrate. This technique permits somewhat less precision to be used in the manufacture of the interference coatings, but it results in high absorption of the transmitted beam. An additional disadvantage is that these compound filters are both heavier and bulkier than an interference filter alone.

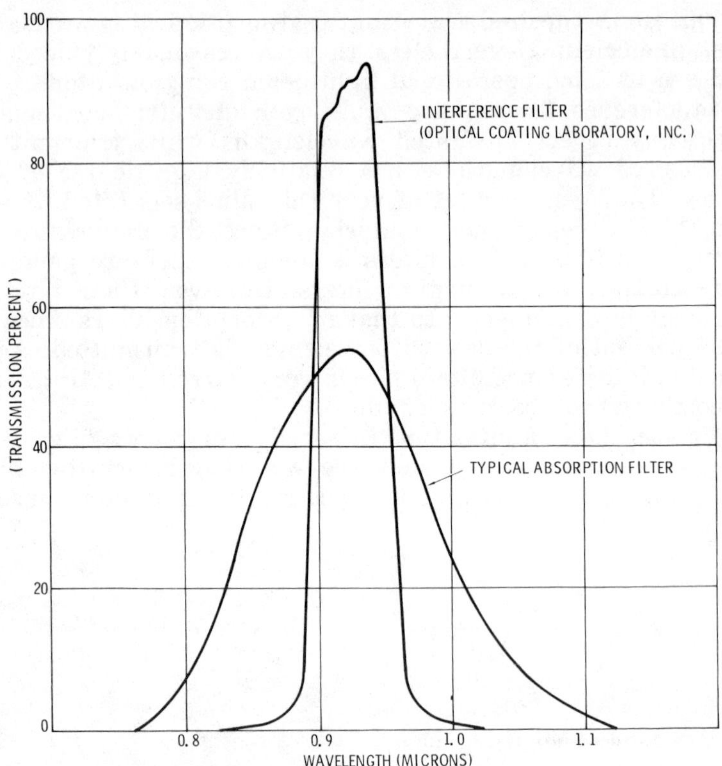

Fig. 5-9. Interference and absorption filter spectral transmission.

MIRRORS AND BEAM SPLITTERS

Plane mirrors are sometimes used to fold the optical system of a light-beam communicator. Folding permits the overall length of the optical system to be significantly reduced, and for this reason folded optics are also employed in some telescopes and high-power lasers.

First-surface mirrors are coated with a highly reflective film of aluminum, silver, or gold on their front surface to avoid undesirable secondary reflections from the mirror's glass front. Second-surface mirrors are coated on the back side of the glass substrate. This permits the delicate coating to be protected by a layer of varnish, but the secondary reflections from second-surface mirrors may be undesirable in some applications.

Most mirrors are made with a glass substrate, but plastic is also being used. Some mirrors are made by polishing a metal substrate to a high gloss.

The beam splitter is a plane mirror which simultaneously reflects and transmits a light beam. As shown in Fig. 5-10A, when the beam splitter is placed at an angle to an oncoming light beam, some of the beam is reflected from each surface. The remainder of the beam passes through the splitter.

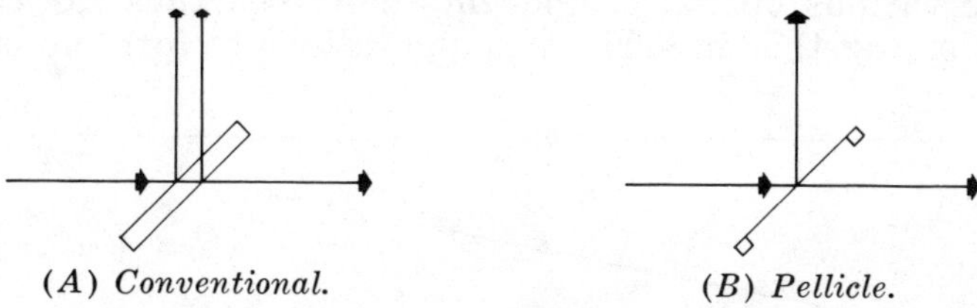

(A) Conventional.   (B) Pellicle.

Fig. 5-10. Operation of a conventional beam splitter and pellicle beam splitter.

An ordinary glass slide can be used as a beam splitter which reflects from 8 to 10 percent of an oncoming beam while passing the rest. Special coatings can be applied to one side of the glass to increase the amount of reflected light. To avoid undesirable secondary reflections from the back surface of a glass beam splitter, an ultrathin splitter called a *pellicle* can be employed. The pellicle is so thin that front and back surface reflections are superimposed for most practical purposes (Fig. 5-10B). Coated pellicles having a wide range of reflection-transmission ratios are available.

A typical application of a beam splitter is its use in a coaxial optical system, such as the one shown in Fig. 5-11. Here, half the light emitted by a source passes through the pellicle and is collimated by a positive lens. Incoming light from a distant transmitter passes through the same lens, and half of the collected radiation is reflected by the pellicle to a detector.

The use of the beam splitter in Fig. 5-11 causes considerable loss in both the transmitted and the received beam. This

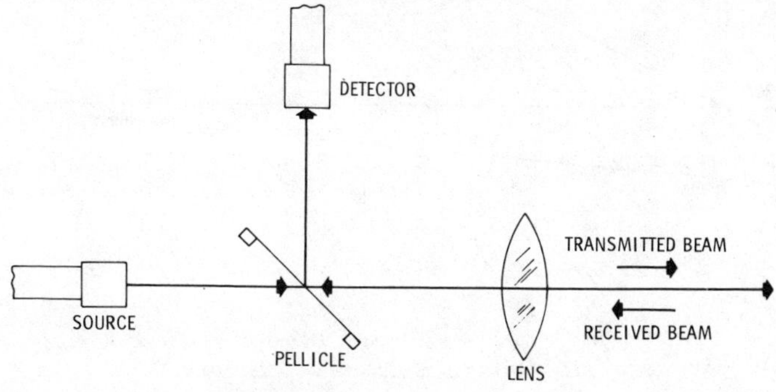

Fig. 5-11. Coaxial optical-communicator transceiver-receiver optics.

efficiency loss, however, can be offset in some applications by the advantages of a coaxial optical system.

OPTICAL SYSTEMS

The various optical components described thus far can be operated together in various configurations to form an optical

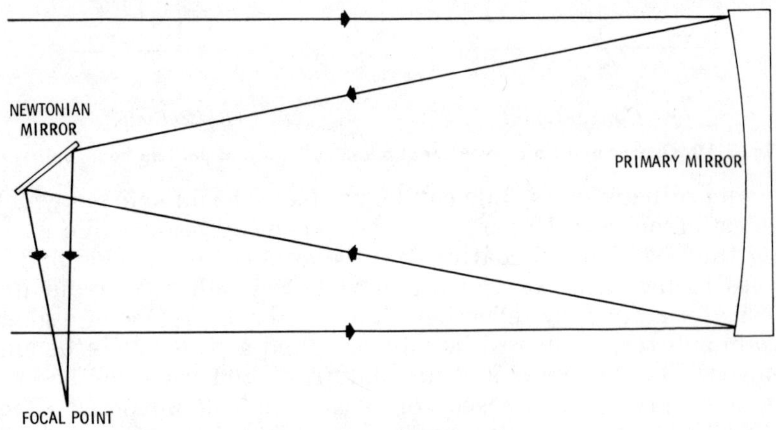

(A) *Newtonion telescope.*

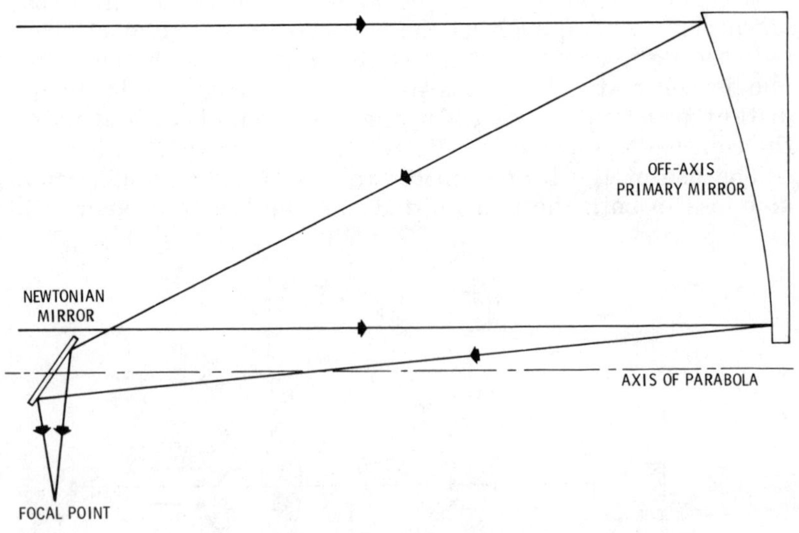

(B) *Off-axis Newtonian telescope.*

Fig. 5-12. Several

system. For example, a compound lens forms a simple optical system.

The telescope is the most useful system for optical communications. Fig. 5-12 shows several telescope configurations that can be applied in optical communications. The Newtonian telescope is the most common configuration for small astronomical reflector telescopes. The telescope consists of a long-focal-length parabolic reflector and a small-plane mirror near the focal point and at a 45-degree angle with respect to the parabola's axis.

A variation of the Newtonian telescope incorporates an off-axis segment of a parabolic reflector. This configuration permits more light to be collected by the parabola since the small Newtonian mirror does not block any of the parabola.

The Cassegrainian telescope incorporates a parabolic primary reflector with a central hole and a small, convex secondary mirror mounted at the axis of the parabola. This configuration is used in many electro-optical systems because the focal region is readily accessible.

Fig. 5-13 is a photograph of a relatively large telescope developed by ITT for a proposed deep-space laser-communications system. It is hoped this antenna will assist in the transmission of up to 30 megabits per second between earth and a satellite 35,400 km (22,000 miles) in space.

The Mangin mirror is a spherical reflector paired with a glass negative lens which corrects the distortions that are characteristic of the reflector. A practical Mangin mirror is simply a negative lens that is aluminized or silvered on its second surface (Fig. 5-14). A protective material is usually placed over the delicate mirror surface. The Mangin mirror is sometimes used in a modified Cassegrainian configuration

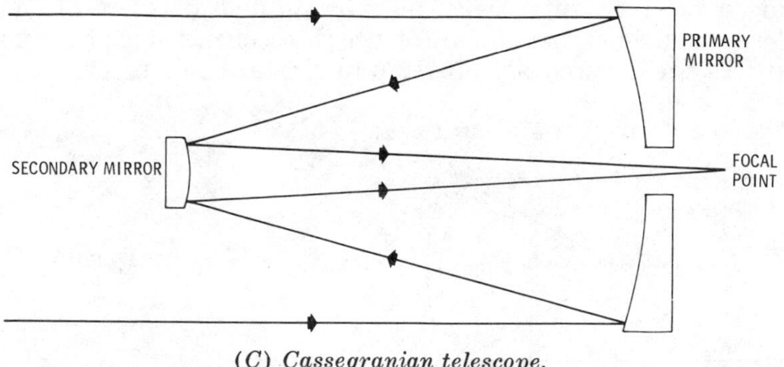

(C) Cassegranian telescope.

telescope configurations.

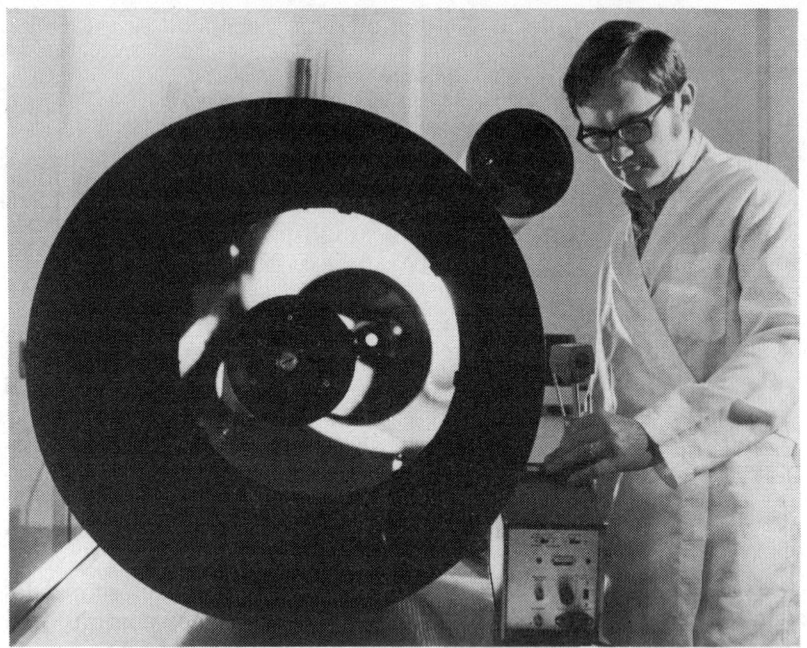

Courtesy ITT

Fig. 5-13. ITT-developed telescope for space-laser communicator.

by employing a small Mangin secondary mirror and drilling a hole in the primary Mangin mirror.

Telescopes, such as those described here, and Mangin mirrors are very useful in moderate to long-range optical-communication links. Frequently, small alignment telescopes are used to aim a communicator toward a distant communicator. Telescopes are also used as collimators of transmitter beams and as receiver antennas. The narrow field of view of most telescopes makes them ideal for use in receivers since they permit less background illumination to strike the detector. Trans-

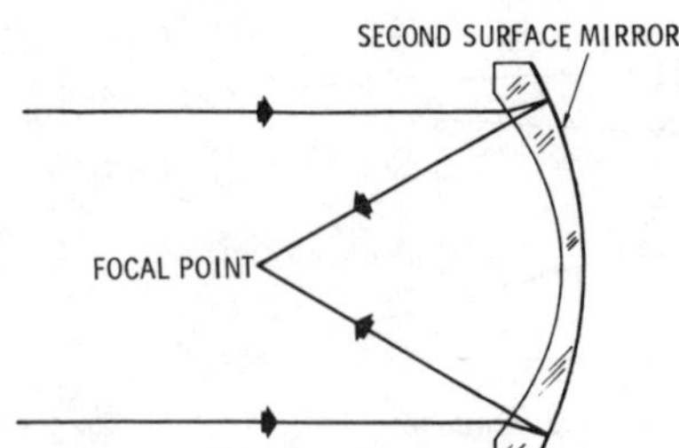

Fig. 5-14. A Mangin mirror.

mitter applications of telescopes generally work best with directional light sources, such as lasers, since their relatively long focal length limits their collection efficiency.

FIBER OPTICS

The current interest in optical communications is largely due to the successful development of very-low-loss optical fibers. In 1870 British physicist John Tyndall found that light will follow a stream of water emerging from a container. In the 1950s, research was begun into the production of glass fibers designed to convey light, and several useful devices and techniques were developed. Among these were fiberscopes, face-plates for cathode-ray tubes, optical couplers for image intensifiers, and others. With the development of low-cost plastic fiber optics in the 1960s, optical fibers could be used to monitor automobile exterior lights, without the need for separate pilot lamps.

Optical communications via optical fibers was an obvious application, but the very high attenuation of early fibers prevented a practical realization of this goal. Transmission losses of 1 dB per meter were typical, and this added up to an intolerable loss factor for a practical communications link. For example, a 100-meter length of fiber would attenuate an incoming signal to about 10^{10} of its initial value. If a maximum of 10 milliwatts could be coupled into such a fiber, only 10^{-12} watts, or 1 picowatt, would reach the other end. The detection limit for the best practical optical receiver is measured in nanowatts, at least three decades *above* this figure.

In 1968 K. C. Kao of Standard Telecommunications Laboratories in Britain found that a slab of high-purity silica glass had a loss equivalent to only 5 dB per kilometer. This major discovery led to a research program for the development of practical fibers having low loss. Corning Glass Works and Bell Telephone Laboratories have led in the production of very-low-loss fibers since 1970, when Corning produced a 16-dB/km fiber. In 1971 Bell Laboratories demonstrated a liquid-core glass fiber with a 13-dB/km loss. In 1972 Corning again took the lead with a 4-dB/km fiber, and in 1973 the firm produced a fiber having a loss of only 2 dB/km. As this is written, Bell Laboratories has just produced a fiber with an incredibly low loss of 1.2 dB/km. This figure is so close to the theoretical transmission limit that future research efforts will be directed toward making such fibers commercially feasible. Routine manufacture of 2-dB fibers may be beyond the capability of the glass production art, but routine manufacture of 10- to 20-dB fibers is well within reach.

The advantages of an optical-fiber communications link are described in Chapter 6. Suffice it to say that after a distance of 600 meters, a 5-dB/km fiber will attenuate only half the light launched into it. This means that repeater systems required to beef up a transmitted signal could be placed farther apart than they are in existing wire-communication links. Of even more significance, the information-carrying potential of a single fiber is significantly greater than that of a copper wire.

Fig. 5-15 illustrates the principle of total internal reflection, which causes a light wave to propagate inside a glass-fiber waveguide. The fiber consists of a central region called the core and an outer layer called the cladding. Since the refractive index of the core is slightly higher than that of the cladding, light striking the core-cladding interface is reflected back into the core. Fig. 1-12 shows light emerging from the ends of a cluster of fibers.

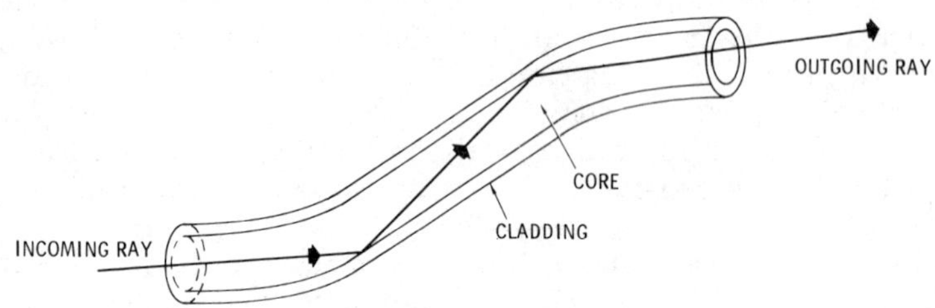

Fig. 5-15. Light transmission in an optical fiber.

Variations of the basic core-cladding fiber can also transmit a light beam. For example, the Nippon Sheet Glass Co., Ltd. of Japan has developed a self-focusing fiber, designated Selfoc, which incorporates a gradual variation in refractive index between the "core" and the cladding. Light rays or modes propagate along a Selfoc fiber in a wavelike, sinusoidal manner instead of in the linear manner characteristic of a fiber with a sharp refractive index boundary at the core-cladding interface.

This gives Selfoc fibers the desirable property of lessened pulse broadening, a phenomenon described later in this chapter. Fig. 5-16 compares light propagation in a conventional fiber with light propagation in a Selfoc fiber with a graded refractive-index boundary.

Bell Laboratories has developed a fiber which uses a single silica glass of constant refractive index. This unique fiber consists of a hollow tube containing a thin rod mounted upon a support plate. Light is carried inside the thin tube, and the outer cylinder serves to protect the inner light-guiding assembly. Fig. 5-17 shows a beam of light emerging from one end of this fiber and illuminating a screen.

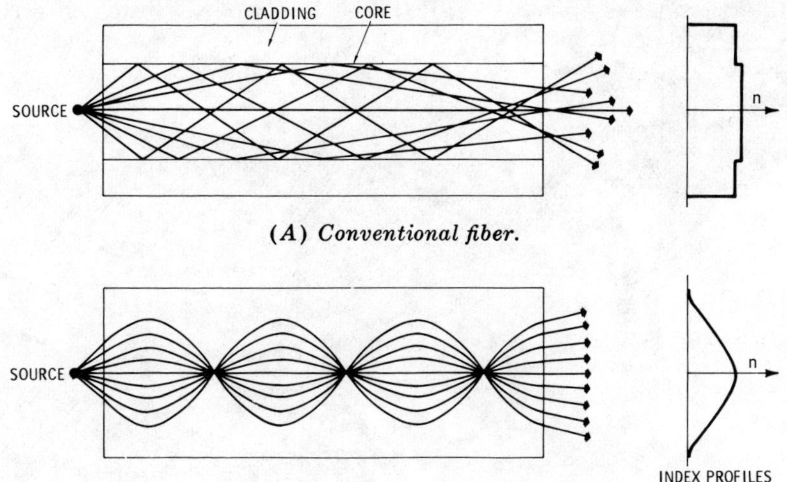

(A) *Conventional fiber.*

(B) *Fiber with graded-index cladding.*

Fig. 5-16. Ray propagation in conventional and graded-index fibers.

In a fiber with parallel sides, light can be launched into a fiber with an acceptance angle governed by the following equation:

$$\sin \theta = \sqrt{n_1^2 - n_2^2} \qquad \text{(Eq. 5-2)}$$

where,

θ is half the acceptance angle,
n_1 is the core index of refraction,
n_2 is the cladding index of refraction.

Equation 5-2 gives a figure of merit for the fiber-acceptance angle. This figure is called the *numerical aperture* (na). The na corresponds to the sine of half the acceptance angle.

The optical radiation density transmitted by a fiber is given by:

$$E = L \ (na)^2 \ T^r \ watt/cm^2 \qquad \text{(Eq. 5-3)}$$

where,

E is the radiation density,
L is the radiance at the fiber input,
T is the transmission of the fiber per unit length,
r is the fiber length.

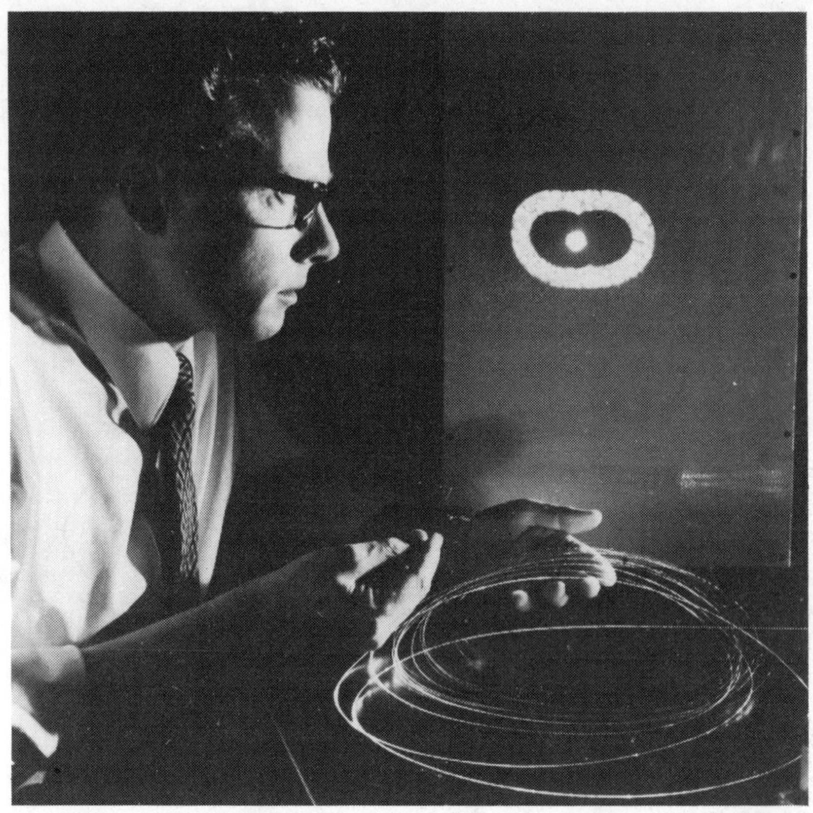

Fig. 5-17. Light beam emerging from high purity optical fiber and illuminating a nearby screen.

A realistic assessment of state-of-the-art LEDs reveals that about 100 microwatts can be launched into an optical fiber having an na of 0.2. This figure can be doubled by doubling the na of the fiber. Stripe-geometry injection lasers may launch a few milliwatts or more into a fiber.

MULTIMODE AND SINGLE-MODE FIBERS

Most of the early glass fibers that were evaluated for possible communication roles had a central core from 50 to 100 microns in diameter. A core this thick permits several thousand light rays to propagate in what are called *modes*.

A characteristic of multimode fibers called *pulse broadening* has resulted in the development of fibers that permit only a single axial mode to propagate. These single-mode fibers

116

have a central core only a few microns or so in diameter. Fig. 5-18 shows mode propagation in single-mode and multimode fibers.

Pulse broadening in multimode fibers has two causes. One is a differential delay caused by the increased path length of all modes other than the lowest-order axial mode. Since the axial mode travels straight through the fiber, it arrives at the de-

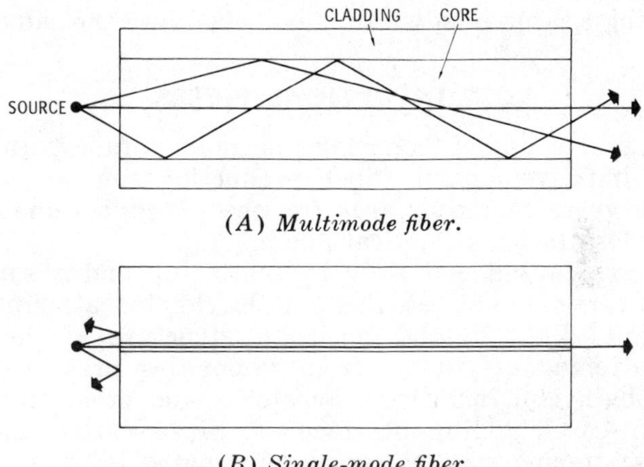

(A) Multimode fiber.

(B) Single-mode fiber.

Fig. 5-18. Mode propagation in optical fibers.

tector before the off-axis modes. As shown in Fig. 5-19, this causes a time distortion of the original pulse. Pulse broadening is also caused by the fact that the speed of light for various wavelengths varies slightly as a function of refractive index. This poses no problem for lasers having a nearly perfect monochromatic output, but other lasers, and LEDs in

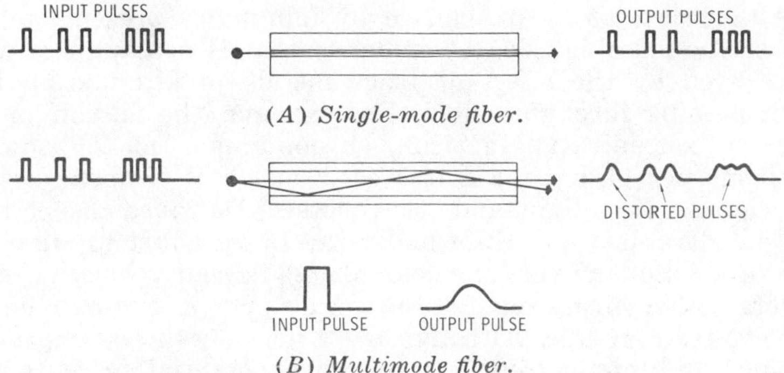

(A) Single-mode fiber.

(B) Multimode fiber.

Fig. 5-19. Pulse distortion in multimode fiber.

particular, may have sufficient spectral width to cause wavelength-induced delay.

Scientists at Bell Laboratories and elsewhere have made extensive studies of pulse broadening. They have found that the wavelength effect upon pulse broadening causes a spread of 3.6 nanoseconds/km in the case of an LED having a spectral bandwidth of 36 nm and emitting into a pure silica fiber. A delay of only 0.2 ns/km occurs when radiation from a GaAs laser having a 2-nm bandwidth is launched into the same fiber.

LOSS IN OPTICAL FIBERS

Low-loss fiber-optics technology is in its infancy, and considerable improvements in fiber production can be expected in coming years. A major area for needed improvement concerns the loss factor in optical fibers.

Losses are caused primarily by scattering and absorption. The most fundamental loss factor is Rayleigh scattering—the scattering of light by the molecular structure of the fiber. Other scattering originates from impurities present in the fiber, bubbles, and inclusions. Scratches and other imperfections at the core-cladding interface also cause scattering. Many of these scattering sources can be eliminated by careful fabrication and materials-processing techniques, but Rayleigh scattering is a fundamental loss factor and sets the limit for scattering-induced loss. Fortunately, Rayleigh scattering occurs as a function of wavelength, with shorter wavelengths being scattered more than longer ones. For example, the theoretical scattering loss at 900 nm is less than 2 dB/km and is only 1 dB/km at 1 micron. These are near the peak output wavelengths of GaAs and Nd:YAG lasers, two excellent light sources for fiber-communication links.

Absorption losses are caused by impurities present in the glass. The ultrahigh degree of purity that is necessary is well illustrated by the effect of trace metals in attenuating the light passing through a fiber. For example, the maximum allowable concentration of iron, chromium, or nickel is only 20 parts per billion for a loss of 20 dB/km. Copper, cobalt, manganese, vanadium, and other metals also cause absorption.

Considerable research is underway in an effort to simplify the production of very-low-loss fibers. Already, Corning and several other firms manufacture high-purity, low-loss fibers for commercial sale. Although costs are now high, improved production techniques will result in substantial price reductions. Ultimately, high-purity optical fibers may be less expensive than copper wire since there is virtually an endless supply of sand—the raw material for glass—and a limited supply of copper.

SPLICING AND TERMINATING FIBERS

Techniques for splicing and terminating copper wires are well developed. Splicing and terminating glass fibers, however, is an entirely new technology. While a broken wire can be spliced with a simple soldering operation, the two ends of a broken glass fiber must be precisely aligned before the joint can be secured. Since the core of a fiber can be only a few microns in diameter, aligning two broken fiber ends can be a difficult procedure.

Several interesting techniques for splicing broken fibers have been invented. One uses epoxy and mechanical clamps to seal the fiber ends together. Probably the best technique is to fuse the broken ends together with heat. Splices made in this manner have shown losses as low as 10 percent.

Whatever splicing technique is utilized, technicians who must perform splicing in the field will have to receive special training, and new tools and splicing equipment will have to be designed. The task of designing reliable tools and techniques for splicing together hair-thin glass fibers is challenging indeed.

Terminating fibers can be accomplished by installing a detector or an emitter in a conventional electrical connector and then by connecting the end of the fiber directly to the detector or the emitter with epoxy. Alternatively, the fiber end can be placed directly against the detector or the emitter by a coupling sleeve. Fig. 5-20 shows methods proposed to couple three promising light sources into an optical fiber. Modifications of these three basic approaches are also being considered. For example, Bell Laboratories has found that GaAs laser light can be efficiently launched into a fiber without the use of a coupling lens by simply forming a spherical end on the fiber with heat. A typical double heterostructure GaAs laser emits a beam with a half angle of 1 radian × 10 milliradians. Only 15 to 16 percent of this radiation can be launched into a flat-ended fiber without an external lens, while 60 to 70 percent of the radiation can be launched into the same fiber with a heat-formed spherical end. The problem of terminating fibers is obviously not trivial, and much work is required before effective and economical fiber connectors will become available.

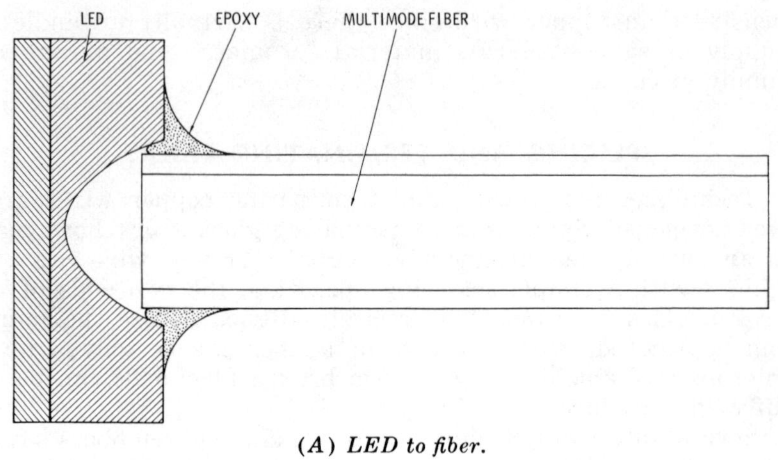

(A) LED to fiber.

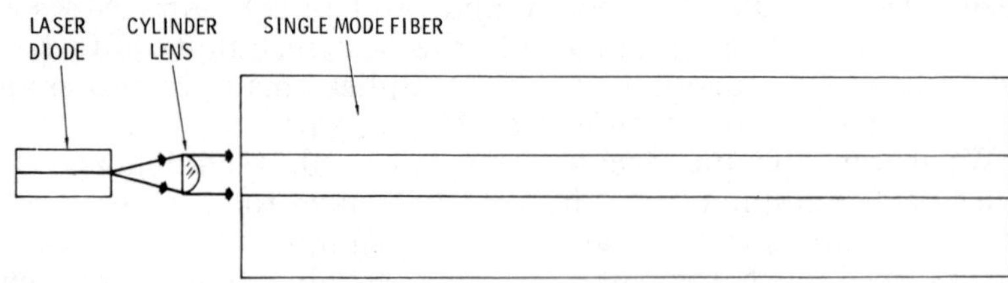

(B) Diode laser to fiber.

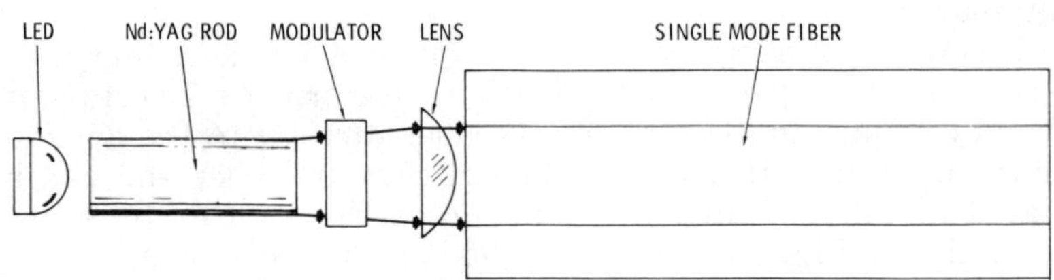

(C) Diode pumped Nd:YAG laser to fiber.

Fig. 5-20. Several source-to-fiber coupling methods.

FIBER MANUFACTURE AND RELIABILITY

Optical fibers can be made by using one of two methods. In the first, a *preform* is constructed from ultrapure glass and is drawn into a fiber. The preform consists of a hollow cylinder (the cladding) into which a rod (the core) is inserted. The fiber drawn from the preform reproduces the preform shape in miniature.

Fig. 5-21 shows glass fibers in several stages of production. Raw material for glass is shown in the lower portion of the

120

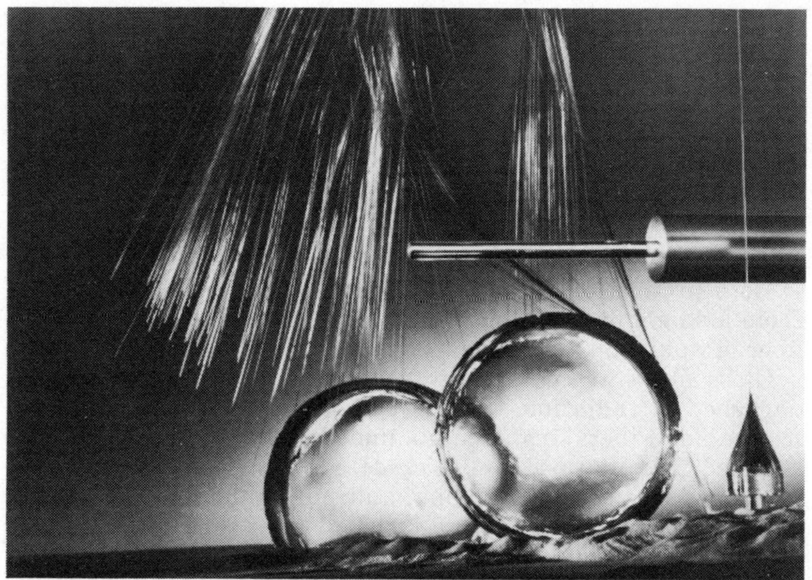

Fig. 5-21. Various stages in the production of glass fibers.

photograph. The two large discs are high-purity glass from which preforms are made. The cylinder of glass emerging into the side of the photograph is a preform assembly, and below

Fig. 5-22. Preform for a single material glass fiber.

that is a preform being drawn into a fiber (the cone-shaped object). The finished fibers are shown at the top of the photograph. Fig. 5-22 shows a preform used to manufacture the single-material optical fiber developed by Bell Laboratories.

The second fiber-production technique requires a concentric crucible. The core glass is melted in the central crucible, and the cladding glass is melted in an outer crucible which surrounds the central one. Both crucibles open at a common, concentric aperture from which a fiber is formed.

Both production techniques can produce kilometers of fiber from a single preform or a batch of molten glass. The finished fiber is wound on a drum, very much like copper wire is wound.

Glass fibers are potentially very reliable. Although glass is darkened by radiation, the effect of cosmic radiation should be negligible. Fibers installed in underground conduits will be shielded from natural radiation sources.

Glass fibers are very strong, and pure silica has a strength of 2×10^6 pounds per square inch—several times the strength of steel. But surface imperfections caused by abrasion and corrosion can cause brittleness and even mechanical failure. Outer sheathings and coatings may help prevent fiber fatigue.

CHAPTER 6

Optical-Communication Systems

An almost bewildering variety of light-beam communication systems has appeared since the advent of the laser and efficient infrared-emitting diodes in 1960. While many of these systems were laboratory breadboards designed to verify modulation efficiency, detector sensitivity, and other communication parameters, many were assembled into working communicators and tested during actual operating conditions.

This chapter describes many of the experimental and operational optical-communication systems which have been developed since 1960. Before describing actual systems, a discussion of system design is important since the amazing variety of optical communicators is a result of many design tradeoffs.

SYSTEM DESIGN

The design of an optical-communication system, particularly one intended for communications through the atmosphere, involves a great many tradeoffs. For example, an infrared-emitting Nd:YAG laser is a desirable optical source because of its high efficiency, but a less-efficient frequency-doubled Nd:YAG laser is more desirable since its output wavelength can be better detected by sensitive photomultiplier tubes. Numerous other tradeoff considerations exist. A narrow transmitter beam helps ensure privacy and long range, but pointing problems associated with narrow-beam systems are significant. Photomultiplier detectors are exceptionally sensitive, but semiconductor detectors are less fragile and require less operating voltage. A narrow-bandpass optical filter reduces the effect of stray light on a detector, but it also attenuates the primary optical signal.

The list of tradeoffs is virtually endless and goes on to include such considerations as source-detector maximum output to minimum detectable signal ratios, optics, packaging, cost, laser safety, etc. There is not enough space in this book to discuss all possible tradeoffs, so a single detailed tradeoff will be presented as a representative example.

LED-INJECTION LASER TRADEOFFS

Both LEDs and injection lasers are popular infrared sources for optical communicators. Since the injection laser produces a higher peak power than the LED, it would therefore appear to be a superior source in a communications application; but, the question of which is superior is far more complex than this. The following discussion illustrates a typical optical-communications tradeoff procedure and includes a detailed consideration of the merits of an injection laser and an LED in a pulse-modulated communicator.

INJECTION LASER

Both the LED and the injection laser have a high power-conversion efficiency (a few percent or more). Certain specially structured LEDs have about twice the efficiency of injection lasers. These diodes, however, are costly, and in most applications injection lasers generally offer higher power-conversion efficiency. Consider, for example, a typical commercial injection laser capable of delivering 4.5 watts when driven with a 10-ampere pulse, and an efficient and inexpensive LED capable of delivering 6 milliwatts at 100-milliamperes bias. Representative commercial devices with these specifications are, respectively, the RCA SG2002 laser and the General Electric SSL-55C LED.

When driven with a 10-ampere pulse, the LED will deliver about 240 milliwatts, or only about 5.3 percent of the 4.5 watts emitted by the laser for the same peak forward current. Equally significant, the LED's power-conversion efficiency has dropped from about 5 percent at 100-milliamperes forward bias to only 1.7 percent at 10 amperes. The power-conversion efficiency of the laser is an impressive 5.6 percent.

In addition to high peak power and conversion efficiency, the infrared emerging from an injection laser is far easier to collimate than that from an LED. This is because the injection-laser emission is directional and emerges from a region on the laser chip only about 0.002 mm × 0.076 mm. Radiation emerging from a source this small can easily be collimated with a very small f/1 lens into nearly parallel light.

Depending on the mechanical structure of the LED, the same lens that produces a highly collimated beam for an injection laser gives a minimum divergence of about 3 degrees when used with an LED. Spot size in the far-infrared field is consequently larger, and power density at the receiver is low.

Thus far, the laser appears to be a superior radiation source. However, its driving requirements lessen its overall advantages. To fulfill the requirement for sustained laser oscillations—an inversion in the population of stimulated electrons—the injection laser must be pulsed with current levels as high as 100,000 amperes/cm^2. Recently, low-threshold double-heterostructure devices have been produced which have reduced required driving current to only a few thousand amperes per cm^2, but these devices have operating lifetimes of only a few hundred hours when operated at a repetition rate of about 1 kHz. Practical operational current densities for good-quality single-heterostructure injection lasers range from 40,000 to 50,000 A/cm^2. In short, to achieve the population inversion prerequisite to laser action, very high current densities must be present in the semiconductor junction.

In order to reduce the current to practical levels of about 10 to 100 amperes, injection lasers are made quite small. The RCA SG2002 described here, for example, has a junction area of only 0.076 mm × 0.305 mm. This reduces the required current to about 10 amperes. But, if this current is applied continuously, the tiny volume of the laser cannot sink it without being destroyed by thermal action. For this reason conventional nondouble-heterostructure injection lasers must be operated at pulse widths of typically no more than 200 nanoseconds. To avoid a gradual pulse-to-pulse thermal accumulation, duty cycle (total "on" time) must not exceed 0.1 percent.

When applied to the operating considerations of an optical communicator, these driving requirements mean that great care must be exercised in the design of the injection-laser driving circuitry. For most practical considerations, two options are available to obtain the necessary high current and fast-pulse width. One is to switch a high voltage through a controlled semiconductor switch, such as a semiconductor controlled rectifier (SCR). The other is to switch a voltage through an avalanche semiconductor switch, such as a selected transistor or a four-layer diode. In both cases the voltage is stored in a small capacitor until it is discharged through the semiconductor switch and laser.

The SCR technique has the advantage of simple design and operation, but typical SCRs present a relatively high impedance to the very fast pulse required to operate a laser. For this reason several hundred volts are required to obtain the 20 or 30 amperes required to drive conventional single-heterostructure lasers.

Avalanche-transistor circuits can be designed to produce similar current levels for significantly smaller voltage levels. This is because avalanche transistors possess a much lower impedance at fast-pulse widths. In a comparison of practical SCR and avalanche-transistor driving circuits conducted by the author, typical SCRs possessed an impedance of nearly 5 ohms for a 75-nanosecond pulse. Typical avalanche-transistor impedance was only a third this value.

Obviously, before selecting a final circuit, one should carefully study the tradeoffs in diode-laser pulsers. Even if great care is exercised in circuit design, the laser circuit will consume more power than a similar LED circuit.

Injection lasers present other special problems in the design of an optical communicator. For example, the fast-pulse width of an injection laser imposes special design requirements on the receiver, and for most practical considerations only reverse-biased pin photodiodes or avalanche photodetectors coupled into a wideband amplifier will make a practical receiver combination. Also, conventional lasers have a limited lifetime, and their degradation is a continual and irreversible phenomenon. Good-quality single-heterostructure lasers have improved lifetimes, and a study by RCA has shown that lasers operated at 1 kHz possess 80 percent of their original output after 1000 hours of operation.

Other disadvantages of injection lasers include temperature sensitivity and beam-diffraction patterns. Narrow-beam laser systems will not be troubled by the latter phenomenon, but wide-beam laser systems generally exhibit highly irregular power distribution in the emitted beam. Infrared photographs vividly demonstrate this phenomenon. Temperature sensitivity is more serious since it directly affects the output of the laser and possibly its degradation rate. Typical injection lasers exhibit significantly lowered current-threshold values as temperature drops. Therefore, a system designed to achieve maximum power output by operating a laser at, or close to, its maximum current level will fail if the laser's temperature is allowed to drop to such an extent that its maximum permissible driving-current level falls below that delivered by the driving circuit. Depending on the temperature change, the laser may be degraded or even destroyed. In a typical system in which the laser is operated at its maximum current rating, a drop in temperature of only 10 degrees Fahrenheit causes laser failure.

Temperature regulation can be employed to protect the laser, but a simpler procedure is to limit the peak current to a safe value for a specified temperature range. The laser simply would not be operated when ambient temperatures exceeded the allowable range.

Other potential difficulties with injection lasers include electromagnetic pulse (EMP) effects and safety. The EMP phenomenon results from the very fast, high-current discharges required to drive a laser. The EMP effects can cause substantial interference in nearby electromagnetic receivers, such as radios and sensitive aircraft navigation equipment. Fortunately the effects are generally minor. In a test conducted by the author, an injection-laser system produced no discernible interference in an a-m radio receiver so long as the two were separated by at least 75 cm.

Safety is a less precise area. The high voltage (i.e., several hundred volts) necessary to operate some laser communicators poses a potential hazard, but only if the laser's protective housing is opened. Of more concern is the potential ocular hazard to persons within the laser's field of view.

At the outset of any discussion on laser ocular safety, it is essential to note that this topic is controversial. The laser cane (an aid for the blind) has been independently evaluated for safety by five separate institutions. The evaluations included exposure tests with laboratory primates, and no discernible retinal deterioration was noted. These tests were conducted under worst-case conditions, with the animal being exposed in one instance for 20 seconds, and in another for 30 minutes. Both times are far longer than the likely exposure times for a person whose eye or eyes happen to momentarily fall within one of the cane's three beams.

Nevertheless, recent reports indicate laser-damage threshold levels may be far lower than previously expected. For this reason and others, the Bureau of Radiological Health of the Department of Health, Education, and Welfare has proposed stringent safety guidelines for most laser products manufactured, sold, and operated in the United States. Optical communicator systems are not included in the present standards, but Wilbur F. Van Pelt of the Electro-Optics Branch of the Bureau's Electronic Product Division recently stated that safety requirements for laser ranging systems and communicators will be implemented should their use "become well defined and pose a significant hazard to the general population."

This is not the place for a detailed discussion on the implications of potential injection-laser hazards and government regulations. It is likely, however, that the new government regulations will expand the current controversy. Numerous references and detailed discussions on laser safety are available elsewhere.

LEDs

Although the injection laser is superior to the LED in power output and ease of beam collimation, the laser's driving requirements are a significant drawback. The LED becomes a viable, alternative optical source for this reason.

A LED can deliver 100 milliwatts from a 3-ampere current pulse. This represents only about 2.2 percent of the RCA SG2002 laser's peak power, but the LED pulse can be significantly wider than the laser's. A 5-microsecond pulse is far easier to detect and amplify than one only 100 nanoseconds wide.

Of even more significance is the simplicity and compact size of the circuitry required to generate 3-ampere pulses for an LED. A simple regenerative amplifier with but four components and powered by a miniature 9-volt battery will easily fulfill the pulse requirements.

LEDs possess other advantages over injection lasers: they are very economical, a great variety of commercial devices are readily available, and operating lifetimes are orders of magnitude greater than those of injection lasers.

This discussion of the relative merits of injection lasers and LEDs is summarized in Table 6-1. A careful study of these considerations and of desired range capability and receiver sensitivity is necessary before a choice can be made between these two optical sources.

OPTICAL LINKS VERSUS EXISTING SYSTEMS

The ultimate tradeoff is whether the cost and the capability of an optical link warrant replacement of an existing system.

Table 6-1. Injection Lasers Versus Light Emitting Diodes

Characteristic	Laser	LED
Peak Power	100 W	300 mW
Average Power	15 mW	250 mW
Power Efficiency	6%	10%
Maximum Duty Cycle	0.1%*	100%
Rise Time	1-50 ns	1-300 ns
Beam Divergence	20°	10°-180°
Spectral Width	3.5 nm	40 nm

* Double-heterojunction devices will operate cw at room temperature.
NOTE: These characteristics are representative and may vary significantly.

Martin Marietta Aerospace recently completed a study which compared the feasibility of using optical and millimeter wave links for Air Force intrabase communications. The study concluded that for existing technology, optical and millimeter wave-data links have essentially equal capability over an atmospheric range of 1-3 km. Below 1 km, optical links are superior; beyond 3 km, millimeter wave links are best. The study concluded that LEDs are the preferred optical source for atmospheric ranges up to 1 km. Injection lasers are best for 1-3 km, and CO_2 and Nd:YAG are required for longer-range links. These results apply to atmospheric links, and glass-fiber systems would probably surpass conventional wire links in most cases.

Deep-space communications is another area where tradeoffs have been conducted between optical links employing lasers and existing microwave systems. In one study conducted by Bell Laboratories in anticipation of the "Grand Tour"—a proposed space-probe flyby of Jupiter, Saturn, Uranus, and Neptune scheduled for 1977—it was concluded that microwave deep-space links are somewhat superior to laser links. The study concluded that the primary drawback to the laser approach was the requirement of an airborne or satellite receiving station to relay signals to the ground when cloud cover was present. The Bell study identified several other problem areas for the laser approach, particularly in the design of optics for the transmitter antenna, but noted that emerging technology might eventually change the picture in favor of the laser.

The Bell study was completed in the late 1960s, and new laser developments, particularly in the field of high-power CO_2 lasers, have greatly improved the prospects for laser deep-space data links. The exceptionally high data-rate capability of the laser over the microwave link is particularly attractive.

OPTICAL COMMUNICATORS SINCE 1960

The remainder of this chapter describes many of the experimental and operational light-beam communication systems that have appeared since 1960 and the advent of the laser and the efficient infrared-emitting diodes. The systems are categorized under four general headings according to range and transmission medium: very short range, medium range, space links, and optical waveguide links. Of course, many systems could be classified under more than one heading, so classification is according to primary application.

SHORT-RANGE SYSTEMS

For the purpose of this discussion, a short-range optical-communications link is one which requires no auxiliary optical system. For example, a system that employs an infrared-emitting diode to transmit information about the temperature of a cryogenic Dewar to a silicon phototransistor a few centimeters away is considered to be a short-range optical-communications system.

Literally dozens of commercial short-range optical-communications links have been demonstrated, and many are in operation. One of these is an optoelectronic coupler, a light source paired with a light detector. A typical coupler, which is also called a source-sensor pair or an optoisolator, consists of an infrared-emitting diode and a silicon photodiode or phototransistor mounted in a metal, ceramic, or plastic package. Fig. 6-1 shows a typical source-sensor pair.

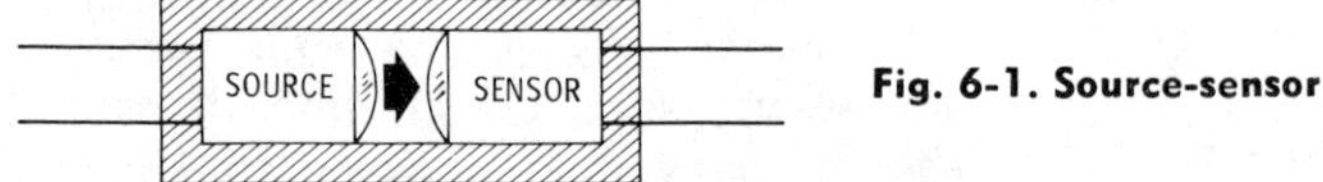

Fig. 6-1. Source-sensor pair.

Optoelectronic couplers are frequently used to electrically isolate one circuit from another while providing a convenient means for data transfer. Since couplers can isolate potential differences of up to several million volts, a number of unusual applications are available.

MEDIUM-RANGE SYSTEMS

For the purpose of this discussion, a medium-range optical link has an atmospheric range of several meters to 100 kilometers or more. Numerous medium-range communicators have been built and tested. They have incorporated an amazing variety of optical sources, detectors, modulation methods, and optical antennas. Most have ranges of several kilometers or more. While many of these medium-range systems are experimental, some are commercially available. Since this category of optical communicators is so broad, this discussion will be subdivided according to the transmitter optical source.

LIGHT EMITTING DIODE SYSTEMS

In 1963 Texas Instruments built several prototype portable optical communicators utilizing high-power GaAs LEDs. These systems utilized 10-cm parabolic reflectors for both transmitter and receiver antennas. The complete transceiver system was handheld, weighed less than 4.5 kilograms, and had a range of several kilometers or more.

Since the Texas Instruments demonstration of a portable GaAs LED communicator, numerous other operational systems employing LEDs have been built. High-speed data transmission between computers and remote terminals is a particularly viable application, and several firms now market LED transceivers for this purpose. One of the first such systems was the Optran (TM), a transceiver employing an infrared-emitting diode. A product of the Computer Transmission Corporation, Optran has a clear-weather range capability of 1.6 kilometers and a data rate of 1200 to 250,000 bits per second. The transceivers are housed in attractive, weather-resistant enclosures measuring 48 cm × 30 cm × 13 cm.

Optran and similar data links manufactured by other firms offer a great deal of flexibility to such computer users as universities, businesses, and military installations. The data rate is much faster than that provided by conventional telephone lines, and the remote terminal can be easily moved by simply moving and realigning its transceiver.

LED links have other applications as well. For example, the Telebeam Corporation uses a 20-milliwatt infrared LED to send closed-circuit pay television to several New York City hotels within a 0.8-kilometer radius. A similar system is used to send market transactions from a New York City stock exchange to a nearby computer facility. And a cable television firm in Boulder, Colorado uses an LED transmitter to send video signals from a mountaintop antenna site to a central distribution point.

Experimenters are particularly intrigued by the prospects of communicating with relatively inexpensive LED communicators. Several firms have even marketed simple LED optical communicators, and one such unit is shown in Fig. 6-2. This simple circuit has a range of only a few centimeters, but its simplicity permits even a very young experimenter to gain an appreciation for some of the more basic aspects of optical communications.

Construction details for the Opticom, an experimental optical communicator with a 300-meter (1000-foot) range, appeared in *Popular Electronics* magazine.[1] Several hundred in-

Fig. 6-2. Author's son Eric operating a simple commercial light-beam communicator.

dividuals and a number of military and commercial groups assembled working versions of the communicator. Since the Opticom article appeared, several other LED communicators have been described in electronics magazines. All of these are described in *Light-Beam Communication Circuits*, a companion volume to this book by the same author and published by Howard W. Sams and Company, Inc.

---

[1] Forrest M. Mims III and Henry E. Roberts, "The Opticom," *Popular Electronics*, November 1970, pp. 45-50.

132

EARLY INJECTION LASER SYSTEMS

Numerous communicators employing injection lasers have been developed. Since the injection laser can be modulated by simply varying the amplitude, rate, width, or position of the driving pulses, very compact and relatively simple laser communicators are possible. The advent of continuously operated injection lasers in 1970 made possible even simpler modulator circuits.

IBM, RCA, and Korad developed some of the earliest injection-laser communication systems. Fig. 6-3 is a block diagram

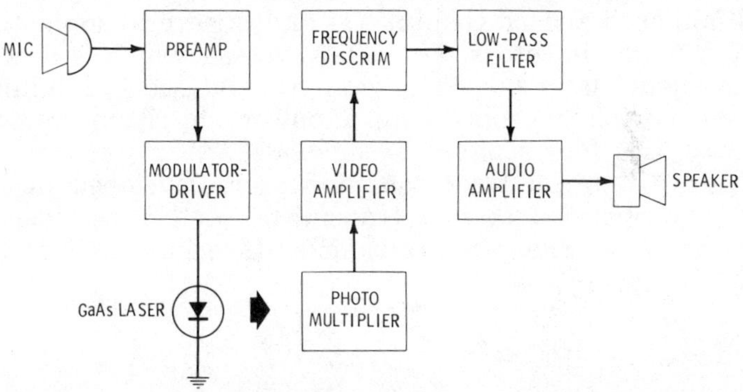

Fig. 6-3. IBM injection-laser communicator.

of an IBM audio communicator developed in 1963, scarcely a year after the demonstration of the first injection laser. The IBM system used a GaAs diode cooled to 77°K by a liquid-nitrogen bath. The laser was driven by a 12-kHz stream of 100-nanosecond, 5-ampere pulses supplied by a four-layer diode relaxation-oscillator circuit. A modulator circuit varied the charging time of the oscillator's discharge network at an audio rate to give a 4-kHz audio bandwidth. The laser emitted 0.2 watt in the near infrared at about 840 nanometers.

The IBM laser-transmitter circuit provided a model for many subsequent pulse-frequency-modulated (pfm) injection-laser transmitters. Experiments with the system and a companion receiver enabled IBM scientists to evalaute the potential of injection-laser communicators in a laboratory environment and to propose the adaptation of GaAs laser systems to satellite links, astronaut-earth communications, and injection-laser radar systems.

The IBM laser communicator required that the laser be cooled with liquid nitrogen—an undesirable feature. In 1964, however, RCA developed an injection laser that could be operated in a pulsed mode at room temperature. This permitted the assembly of truly miniature laser-communication systems, and RCA engineers devised several.

The first such system was assembled within weeks of the introduction of the new laser and, like the IBM system, it employed pfm with a 20-kHz carrier frequency and a 6-kHz audio channel. The laser was pulsed with 30-ampere pulses only 30 nanoseconds in width, and selected lasers provided as much as 1.5-watts peak pulse power at room temperature.

This early RCA transmitter was followed by a series of experimental injection-laser communicators culminating in the GT-7 launched aboard Gemini VII and in several sophisticated military communicators. The GT-7 system, which is described in more detail later in this chapter, was housed in a miniature hand-held package containing a power supply, a modulator circuitry, and four separate laser diodes.

Fig. 6-4 shows a laser communicator, developed by RCA, which incorporates both a transmitter and a receiver in a single housing. The pfm transmitter incorporates two lasers

Fig. 6-4. Miniature laser transcevier.

for a total peak-pulse power of several watts and has a photo-diode and a narrow-bandpass optical filter mounted at the focal point of a 10-cm diameter parabolic reflector. This minia-ture laser transceiver is unique because the receiver mirror and the photodiode assembly both fold down to provide a very compact package. This system has a laser beam width of 8 mr and a maximum range in excess of 5 kilometers. The unit op-erates for 2.5 hours from self-contained batteries.

The success of RCA laser communicators has stimulated the development of a variety of other laser transceivers. Fig. 6-5 shows a commercial laser communicator developed by the

Courtesy Santa Barbara Research Center

Fig. 6-5. Portable injection-laser transceiver sets.

Santa Barbara Research Center. This transceiver uses a single 2-watt injection laser which emits a 5-mr beam. The receiver incorporates an avalanche photodetector having a 64-nanowatt detection capability at a snr of 30. A complete transceiver weighs only 2.5 kilograms and has a maximum range of 10 kilometers. Fig. 6-6 shows one link of this system in opera-tion. During alignment, the laser transmits a series of tone bursts as a marker signal. The operators zero in on one an-other's marker signal and lock their transmitters in a fixed position to begin communications.

More recently, the Santa Barbara Research Center has developed a laser transceiver which resembles a pair of binoculars having three lens tubes (Fig. 6-7). The laser system resembles an LED version developed by another firm in 1968. The LED system, however, had a range of only a kilometer or two, while the laser system has a range of 5 or more kilometers.

Fig. 6-6. Alignment of portable injection-laser communicator.

Holobeam, Inc. has developed several injection-laser transceivers which utilize sophisticated ppm and simpler pfm. The ppm systems were originally developed during a four-month crash program for the Navy in 1969-70.

One version of the ppm transceiver was installed in a helmet and a belt-mounted power and electronics package. This communicator was originally designed to have a range of 16 kilometers with a beam divergence of 1 mr. The Navy, however, was more interested in short-range communications between ships at sea, so the beam was spread to 300 mr with a resultant range of only 75 meters. Spreading the beam greatly facilitates the pointing of two transceivers at one another from ships in a rolling sea. At 75 meters, the beam has a divergence of 23 meters. Approximately 8 watts of infrared are projected

136

by the GaAs injection laser. Reception is enhanced by employing four large-area silicon photodetectors mounted on the front and rear of the helmet and on both earphones. Since no lenses are used with the detectors, reception can occur with one transceiver helmet pointed a full 180 degrees away from the transmitting unit. The photodiodes have an active area of 1.25 cm² and can accept a beam at an angle of 100 degrees.

Fig. 6-8 is a block diagram of the transmitter portion of the Holobeam ppm system. Operation of the circuit is quite complex and involves the generation of an audio-modulated pulse train containing a superimposed series of doublet synchronizing pulses. The synchronizing pulses are necessary for the receiver to decode the ppm audio signals.

A unique feature of the Holobeam transceiver is a thermistor which is attached to the laser and which monitors its tem-

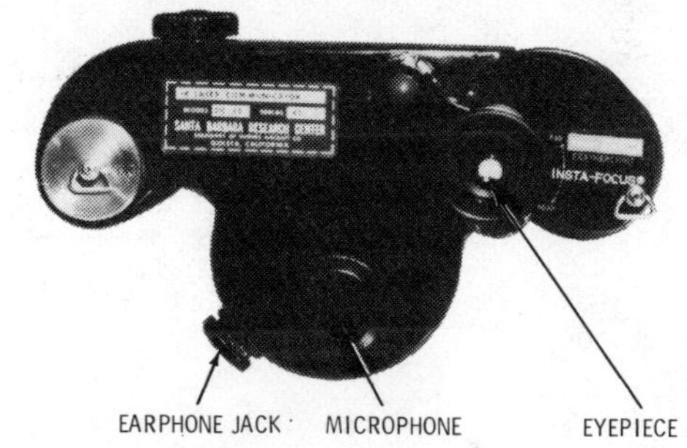

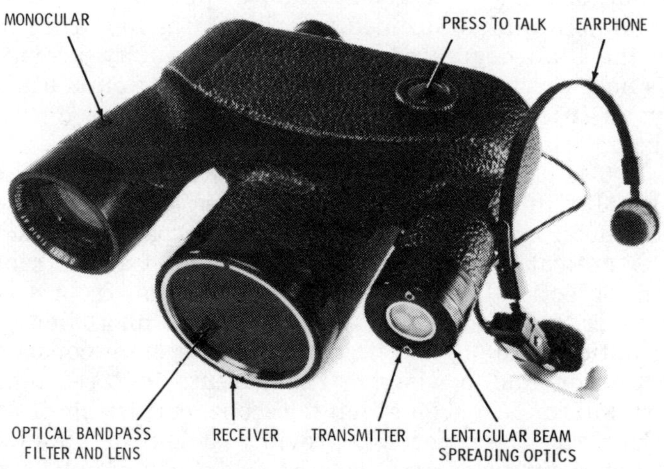

Courtesy Santa Barbara Research Center

Fig. 6-7. Handheld injection-laser communicator.

perature. Since peak laser operating current is governed by temperature, the monitoring circuit permits the laser current to be compensated according to temperature. This ensures optimum operating conditions for the laser and significantly improves the useful operating life.

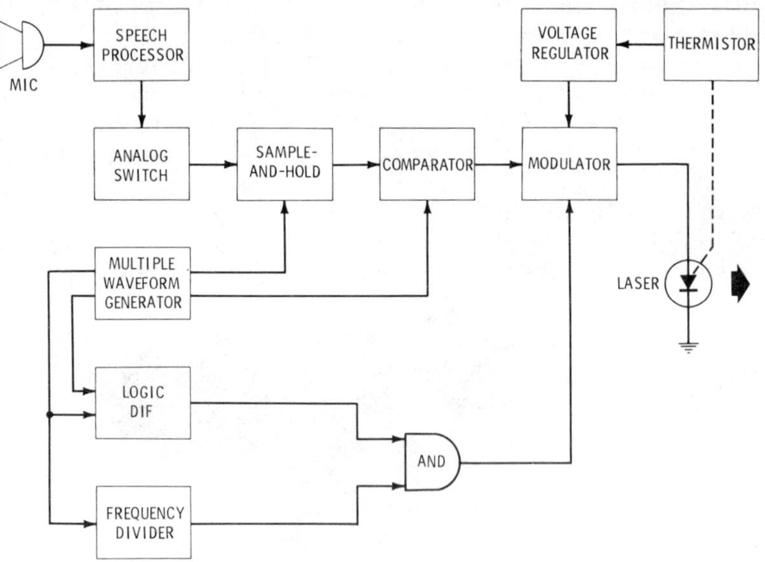

Fig. 6-8. Holobeam ppm injection-laser transmitter.

Because of the cost and complexity of ppm optical communicators, Holobeam has decided to utilize pfm in future models where the high degree of transmission security provided by ppm is not necessary. Pulse-frequency modulation is utilized in nearly all injection-laser communicators because of its simplicity and relatively low cost.

Fig. 6-9 is a diagram of a GaAs laser communicator developed by Martin Marietta Aerospace. This system has a range of 5 kilometers and incorporates a unique coaxial transmitter-receiver optical system. The laser is mounted at the center of a 22-cm objective lens which collects radiation from a remote transmitter and focuses it upon a silicon photodiode via a folded optical system. The folded optical system consists of a plane mirror and a visibly transparent, infrared-reflecting dichroic mirror. An acquisition telescope, consisting of an eyepiece lens and the objective, provides a field of view coaxial with that of the laser and detector to greatly simplify system alignment.

The Martin system was developed to demonstrate the feasibility of air-to-air communications during aerial refueling operations. The system is designed for use with an automatic-tracking mount.

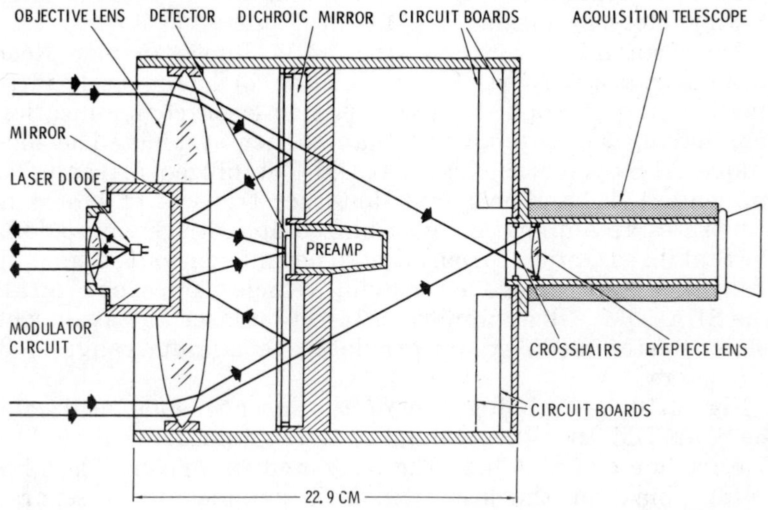

Fig. 6-9. Martin injection-laser transceiver.

CURRENT INJECTION-LASER SYSTEMS

The injection-laser communicators described in the previous section (some of which are still in use) demonstrated the practical nature of injection-laser communicators. As this is written—some eleven years after the first demonstration of a GaAs injection-laser transmitter—a new generation of laser-diode communicators has emerged. These new communicators employ modular laser drivers, sensitive avalanche photodetectors, and new, long-lived GaAs heterojunction lasers. While many of the early laser communicators employing GaAs diodes were limited to specialized military or demonstration applications, the new generation of communicators is intended for practical application in a variety of both military and civilian areas.

An example of the new generation of injection-laser communicators is the SECOM system manufactured by the Data Optics Corporation (1804 Tribute Road, Suite P-205, Sacramento, CA 95815). SECOM has an audio bandwidth from 10 Hz to 15 kHz and a maximum data rate of 150 kHz. The laser is a

GaAs single-heterojunction unit being driven with 10-nanosecond pulses and emitting at least 5-watts peak pulse power. The receiver utilizes a pin photodiode.

The range of the SECOM system is approximately 5 kilometers with a 5-cm diameter receiver lens. A receiver designated COM-14 is available which extends maximum range to about 13 kilometers by means of a 14-cm diameter lens.

American Laser Systems, Inc. (106 James Fowler Road, Santa Barbara Airport, Goleta, CA 93017) has recently introduced a line of sophisticated injection-laser communications gear having a greater range than earlier systems. The most unique ALS system is the SLACOM (*Stabilized LAser COM*municator), a hand-held injection-laser transceiver which incorporates an automatically stabilized optical system made by the Stabilized Optics Corporation to permit continuous tracking of another transceiver from moving vehicles, ships, or aircraft. The SLACOM utilizes miniature injection-laser and avalanche-photodetector modules and provides a maximum range of 32 kilometers.

Fig. 6-10 is a photograph of the laser module utilized in the SLACOM and other communicators manufactured by ALS. The module contains both the laser and its driver. The high-voltage bias for the laser driver is supplied by a separate power pack connected to the module with a miniature cable.

Fig. 6-11 shows a SLACOM unit being operated by an ALS engineer. The unit operates from self-contained batteries which provide up to seven hours of operating time. The laser has a peak power output of 2 watts. Fig. 6-12 shows a complete SLACOM system including headset, belt pack, and hand-carried optical unit.

Optical stabilization is achieved by means of a patented principle which cancels the effects of movement with opposite-phase reflected light. A gyroscope is included to overcome bearing friction and thus to permit stabilization of very-low-vibration frequencies. The complex mechanical nature of the stabilization systems is shown in Fig. 6-13, a view of a SLACOM with its cover removed. The apparatus includes the laser transmitter and receiver modules, sighting telescope, gyroscope, stabilized prism, and batteries. The stabilized pointing system of the SLACOM permits simple hand-held operation over ranges that would require fixed tripod mounts for unstabilized communicators.

American Laser Systems has incorporated its laser and receiver modules into a system having greater range than any existing injection-laser communicator. Designated the Model

736 Laser Communicator, this system consists of a separate transmitter and receiver, each housed in identical weather-resistant anodized-aluminum housings measuring 10 cm × 10 cm × 33 cm. Both transmitter and receiver include a 10×-alignment telescope. Fig. 6-14 shows the 736 transmitter unit, and Fig. 6-15 shows the 736 receiver.

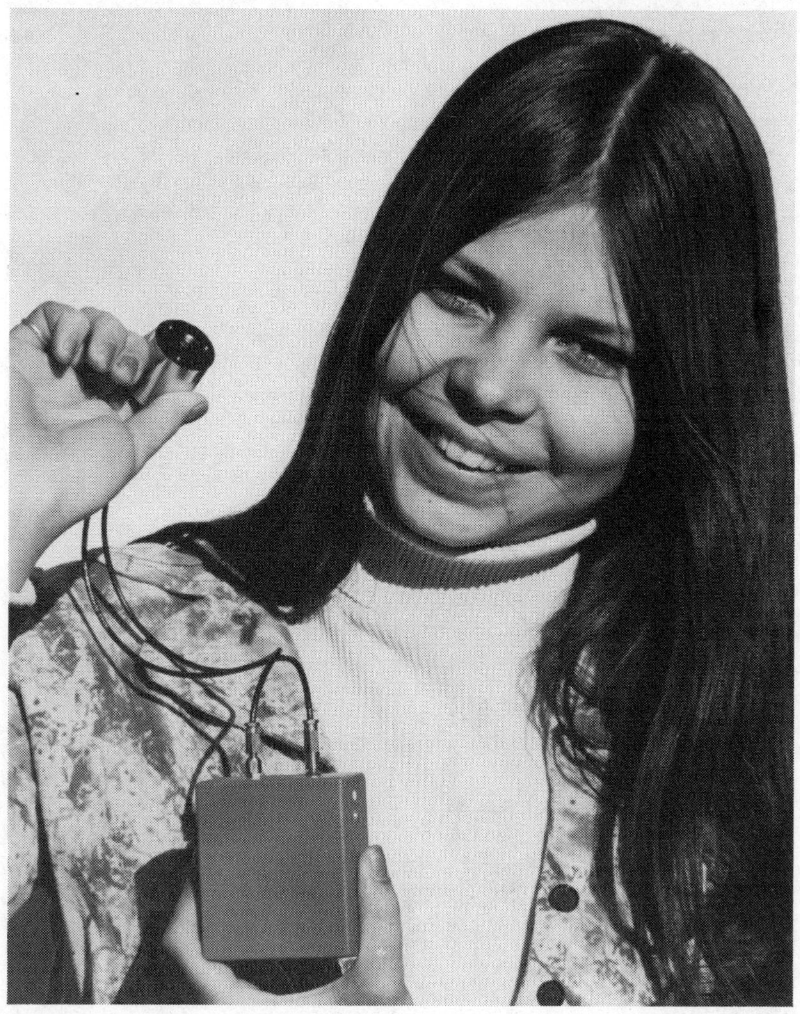

Fig. 6-10. Miniature injection-laser module and power-supply circuit manufactured by American Laser Systems, Inc.

The 736 transmitter employs a single-heterostructure GaAs injection laser with a peak pulse power of 10 watts and with a maximum repetition rate of 10 kHz. A 24-watt laser with a maximum repetition rate of 7.5 kHz is available as an option. Temperature-compensation circuitry is incorporated into the

Fig. 6-11. American Laser Systems SLACOM stabilized injection-laser communicator.

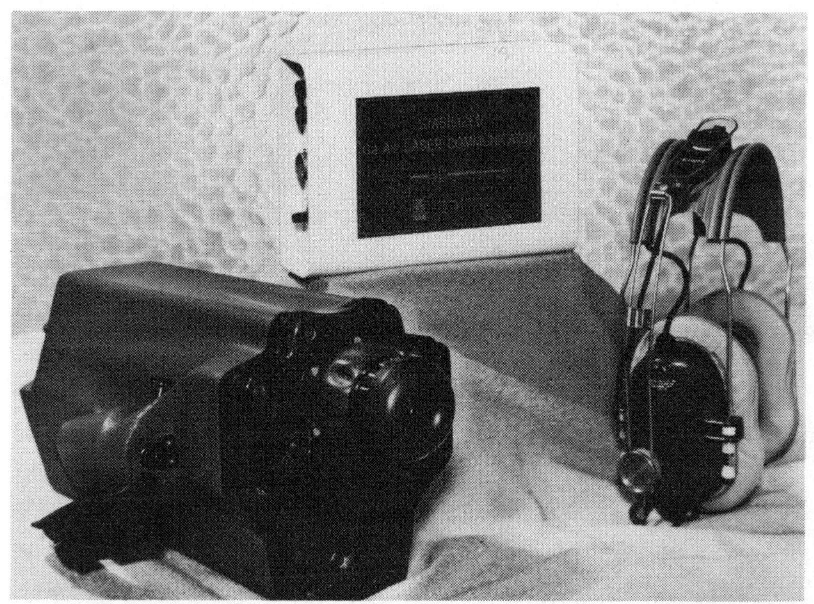

Courtesy American Laser Systems, Inc.

Fig. 6-12. SLACOM handheld unit and belt-mounted power pack.

Courtesy American Laser Systems, Inc.

Fig. 6-13. Internal view of the SLACOM communicator showing complex nature of optical-stabilization systems.

Courtesy American Laser Systems, Inc.

Fig. 6-14. Model 736 laser transmitter.

laser-pulse power supply to maintain constant laser output power from −18°C to 65°C. As in the case of the Holobeam communicator, this significantly prolongs useful laser life.

The 736 receiver contains a silicon avalanche photodiode protected from excessive ambient light by a narrow-bandpass optical filter. The avalanche detector is biased by a temperature-tracking power supply in order to provide optimum sensitivity from −40°C to 60°C. The receiver has a bandwidth

Courtesy American Laser Systems, Inc.

Fig. 6-15. Model 736 laser receiver.

of 15 MHz and can detect an optical signal of only 0.58 nanowatt at this frequency. This is far superior to the sensitivity of pin diodes at similar frequencies.

The 736 system has been routinely tested at ranges greater than 25 kilometers in the presence of moderate atmospheric humidity. When visibility is sufficiently high, ranges up to 40 kilometers have been achieved. A maximum range of 5 kilometers can be achieved 99 percent of the time during almost any expected weather condition.

The 736 transmitter and receiver are easily interfaced with a variety of digital and audio signals. A receiver can even be connected to a transmitter to form an effective repeater station for long-range and terrain-avoidance communications. Fig. 6-16 is a block diagram of the complete 736 system.

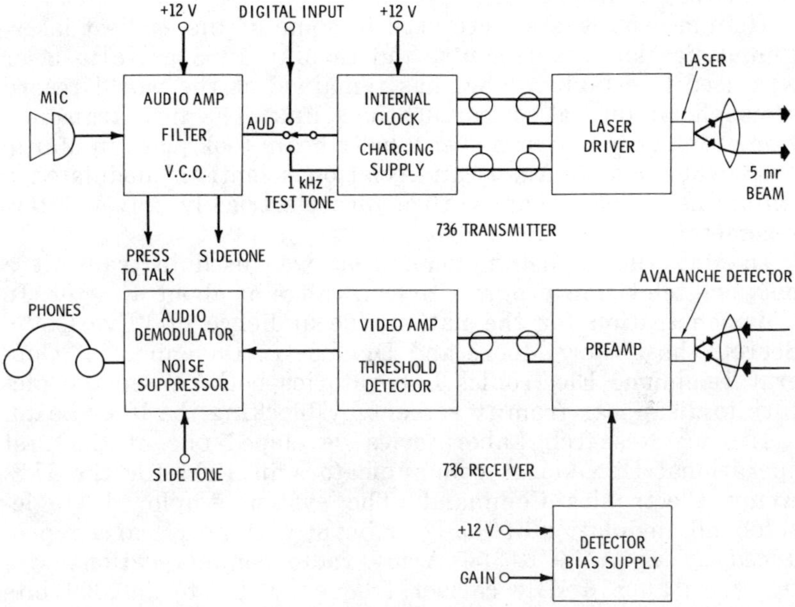

Fig. 6-16. Block diagram of the American Laser Systems 736 laser communicator.

Considerable work is underway to develop even more-advanced injection-laser communicators. One example is an Army-sponsored RCA program to develop a superior double-heterojunction GaAs injection laser capable of emitting 3-watts peak pulse power at a rate of 2.5 MHz at room temperature. RCA has already produced lasers emitting in excess of a watt at 2.0 MHz, and some lasers have been operated at 7 MHz.

Fig. 3-28 in Chapter 3 shows the microwave package used for mounting these experimental lasers.

GAS-LASER SYSTEMS

Several gas-laser systems have been utilized in optical-communication experiments (see Chapter 3), but most work centers around helium-neon, carbon dioxide, and argon lasers. These lasers are well developed, and their high degree of coherence permits a great variety of modulation schemes to be employed. Some gas lasers have significantly higher output capability than semiconductor units, which permits greater range capability. However, the simplest gas-laser transmitter is far larger, heavier, and more delicate than almost any semiconductor laser system.

Helium-neon lasers were used in some of the earliest laser-communication experiments, and in May 1963 a NeHe laser was used to establish what has remained as the world record atmospheric optical-communications link. The first transmission of television signals over a laser beam took place in March 1963 when North American Aviation scientists modulated a helium-neon laser beam with a piezoelectrically driven interferometer.

In May 1963 a helium-neon laser was used to transmit a network television program over a range of about a meter. In a demonstration for the nationwide audience of "I've Got a Secret," host Gary Moore and Dr. Lee L. Davenport of General Telephone Electronics Laboratories both caused the picture to disappear from tv screens by blocking the laser beam.

Hughes Research Laboratories developed one of the first operational HeNe-laser communicators in 1965 for the U.S. Army Electronics Command. The system employed single-sideband modulation of the laser beam with an rf carrier produced by an AN/GRC-50 Army radio communications system with an 876-MHz carrier frequency. Up to 250,000 bits of data per second could be transmitted over a 1.6-kilometer course with the 3-milliwatt beam of the HeNe laser. A silicon photodiode was used to detect the modulated beam.

In 1966 pulse-code modulation (pcm) was employed to obtain much higher data rates from a gas-laser transmitter. One such system, developed by ITT Federal Laboratories under a NASA contract, utilized a 5-milliwatt HeNe laser modulated by a variable polarizing crystal of KDP. The other system, which was built by Hughes for NASA, used a 5-watt argon laser also modulated by a variable polarizing KDP crystal.

Both systems gave a 30-megabit data rate. The ITT system was tested extensively over an 8-km path, and the Hughes system over a 6.7-km path. The higher power of the latter system gave a theoretical maximum operating range in space of 24 million km (1.5 million miles).

In 1967, Hughes employed a frequency-shift keying (fsk) format to modulate a HeNe laser operating at 3.39 microns in the infrared. This 2-milliwatt system could transmit up to 50 megabits per second over a range of 1.6 kilometers to an indium-arsenide photodiode. This system was also built for NASA. More recently, Hughes used a similar modulation method to transmit the band of frequencies between 54 to 216 MHz. This corresponds to transmission of all vhf television channels *simultaneously*.

The success of HeNe laser communicators resulted in a novel method for eavesdropping on private conversations. The method, which was popularized in the general press in the mid-1960s, was supposedly invented by the Central Intelligence Agency. A HeNe laser was pointed at a window of a room in which a conversation was to be monitored. Some of the beam was reflected from the glass window and was detected by a sensitive receiver. Since the window moved slightly in response to the sound waves projected by the speakers, the return beam was amplitude modulated with the conversation. The author has tried this technique and found that even reflected sunlight serves as a usable light source.

Much of the recent emphasis on gas-laser communications centers around the carbon dioxide laser. In 1968, both Honeywell and Hughes developed sophisticated CO_2 laser communicators employing optical-heterodyne detection. The Honeywell system, which was built under a NASA contract, provided two voice channels via mechanical modulation of the laser mirror with a piezoelectric crystal. The Hughes system employed a more sophisticated modulation technique using an electro-optic crystal of gallium arsenide *inside* the laser cavity. This system transmitted a 5-MHz television signal over a range of 28.8 km with 1 watt of 10.6-micron radiation from the laser. The Honeywell system was operated over a 5.4-km test range with 5 watts of 10.6-micron radiation.

In 1973 Hughes tested an 8-megabit CO_2 laser communicator over the 32-km path between Malibu and El Segundo, California. The Army tested this system for 1320 hours of continuous operation over a 4.8-km test range.. Only 65 hours were lost because of adverse weather. This gives a 95-percent reliability factor.

These and numerous other communicators have shown the gas laser to be a viable optical source for practical communications applications. Several operating gas-laser communicators are now in existence, one of which is a sophisticated pcm system utilizing pulsed radiation from HeNe lasers. Developed by the Nippon Electric Company of Japan, this system provides two-way communications over the 14-km path between Yokohama and Tamagawa. In addition to the two main terminal sites at each city, three repeated stations are employed. The maximum separation distance is 4.25 km.

The data rate of the Nippon system is an impressive 123.492 MHz. The system gives 99-percent reliability in all weather conditions over a 24-hour period and gives 99.6-percent reliability during business hours. Sixteen 3-milliwatt HeNe lasers are employed. Each is externally modulated by a lithium-tantalate crystal.

Several firms manufacture very-low-cost HeNe lasers capable of being directly modulated by varying the discharge current. The most diversified line of these low-cost laser transmitters is produced by Metrologic Instruments, Inc. (143 Harding Ave., Bellmawr, NJ 08030). Metrologic was founded by Harry C. Knowles in 1969 with the goal of mass-producing reliable, inexpensive HeNe laser tubes. In *Popular Electronics* (May 1970, pp. 27-42), Knowles described how to assemble a simple a-m communicator by using one of his laser tubes, and many experimenters duplicated the project. In 1970, the Smithsonian Institution demonstrated a Metrologic voice communicator in LASER-10, an exhibit commemorating the tenth anniversary of the laser. Spectators were allowed to talk over the bright red HeNe laser beam.

Metrologic now offers a line of modulated lasers far more sophisticated than the early units built by Knowles five years ago. For example, the ML-669 is a 0.8-milliwatt HeNe laser capable of being modulated from 300 Hz to 500 KHz. This laser can transmit both audio and video information. A 1-volt incoming signal provides full 15-percent modulation of the amplitude of the laser beam. The ML-939 is a 2.2-mW version of the ML-669, but with a modulation rate extending to 600 kHz.

Fig. 6-17 is a block diagram of the ML-669 modulation circuitry. The circuitry consists of a preamplifier, an amplifier, and a current modulator. The modulator controls the output of the laser over a range that is nearly linear from less than 0.5 mW to more than 0.8 mW. The range of the 669 is a mile if suitable collimating optics are used with the laser, and if suitable collection optics are used at the receiver.

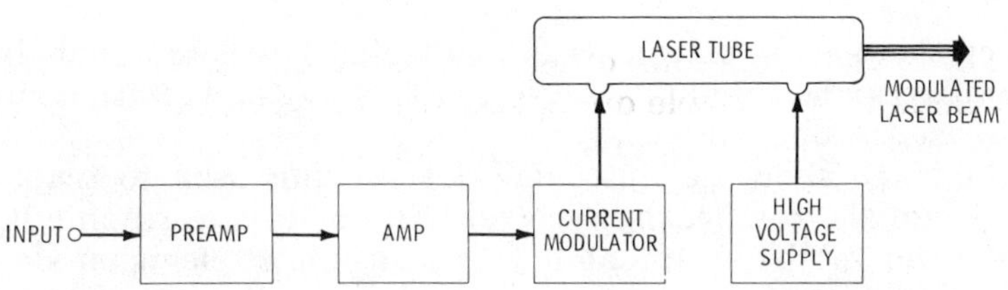

Fig. 6-17. Block diagram of Metrologic ML-669 HeNe laser transmitter.

On-off digital modulation at rates from dc to 160 kHz is provided by the ML-684, a 1-mW NeHe laser. This unit accepts TTL input signals in order to give 100-percent on-off modulation of the laser tube. The tube is designed with a high-transmission-output mirror in order to give a relatively low-gain optical cavity. When the tube current is reduced below threshold level, the gain drops below the optical loss and lasing stops.

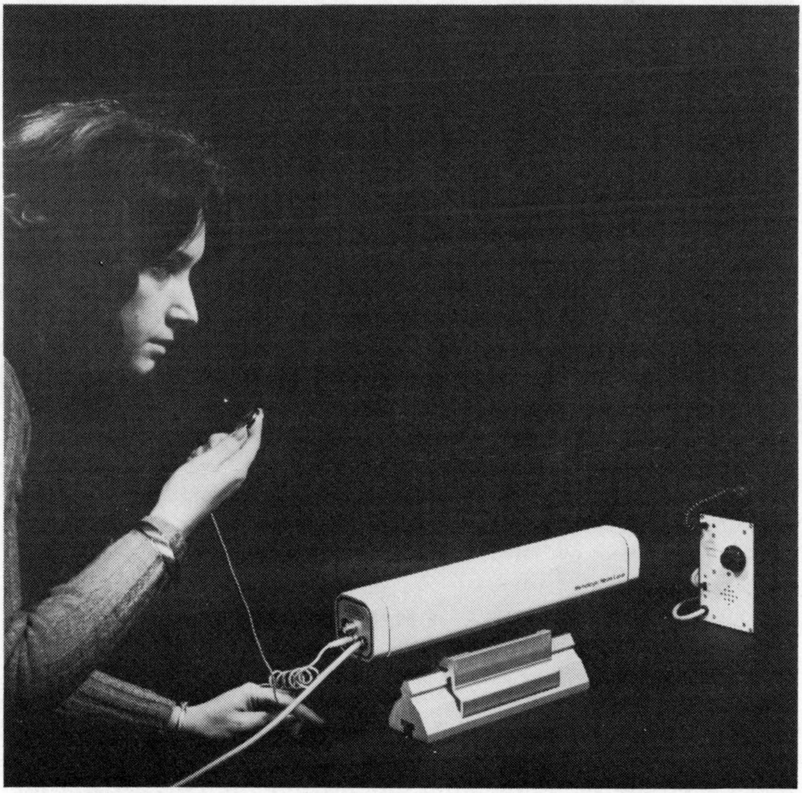

Courtesy Metrologic Instruments, Inc.

Fig. 6-18. Helium-neon laser transmitting voice to nearby receiver.

Since the plasma discharge is unaffected by the current drop, the tube can be rapidly changed to the lasing state with a slight increase in tube current.

Fig. 6-18 shows a Metrologic laser being used to transmit voice signals to a nearby receiver. Fig. 6-19 is a graphic demonstration of the laser's ability to transmit a television signal.

The Metrologic line of modulated lasers is priced from a few hundred dollars to under five hundred dollars. Their reasonable price, coupled with their simplicity of operation and their visible red beam, makes them ideal for use in both demonstrations of laser communications and practical medium-range optical links. Both indoor and outdoor applications for these systems are feasible and have been demonstrated.

Courtesy Metrologic Instruments, Inc.

Fig. 6-19. Metrologic helium-neon video-modulated laser.

SOLID-STATE ION LASER SYSTEMS

Solid-state ion lasers have the very high peak-pulse power necessary for medium-range and deep-space optical communications, but most such lasers are limited in usefulness because of their very low repetition rate. Currently, the best solid-state laser for communication applications is Nd:YAG, a material which gives a 1.06-micron output in the near infrared, with moderately high efficiency. Nd:YAG lasers have produced more than 1000 watts *continuously*, and several hundred watts is relatively easy to obtain. Several Nd:YAG lasers have been externally modulated by utilizing electro-optic crystals. An attractive feature of Nd:YAG is that its invisible output can be converted to a brilliant green light at 530 nm by using certain nonlinear optical crystals to achieve frequency doubling.

Nd:YAG has such a low threshold that semiconductor diodes and lasers emitting at some of its primary absorption wavelengths have been used to pump it to the lasing state with a high degree of efficiency. Chapter 3 describes several such lasers in detail. Later in this chapter, a proposed space-communications application for this laser is described.

Another unique feature of Nd:YAG and some other solid-state ion lasers is that they can be pumped to the lasing state by sunlight. The first sun-pumped laser employed a CaF:Dy crystal which emitted at 2.36 microns when pumped by about 50 watts of sunlight reflected into the crystal by a 25-cm parabolic mirror. Sun-pumped lasers have obvious applications at remote communication sites and in space.

SPACE SYSTEMS

The first laser-communications system to qualify for space operation was an RCA-developed GaAs injection-laser transmitter. The transmitter incorporated four lasers having a total peak-pulse power of 25 watts. It was capable of transmitting a 100-Hz acquisition tone and an 8000-Hz voice channel. The complete transmitter measured only 7.5 × 13 × 20 cm, weighed 2.35 kilograms, and was designed for hand-held operation by an astronaut. The four lasers were aligned with their respective lenses to form a square, approximately 945 meters on a side, from an altitude of 320 kilometers for a beam divergence of 3 milliradians. This gave an allowable pointing error of ±1.5 mr, a reasonable goal considering a good marksman can accurately fire into a ±0.5-mr circle from a prone position.

This injection-laser transmitter, designated GT-7, was carried into earth orbit in 1967 by the Gemini VII astronauts, Frank Borman and James Lovell. Three ground stations were equipped with a receiving telescope and an argon laser. The argon laser was on a tracking pedestal slaved to an FPS-16 radar system designed to lock onto the Gemini's C-band beacon. The argon laser produced a brilliant green (488-nm) beam which was pulsed at 7 Hz and which provided a pointing target for the astronauts.

Unfortunately, tests with the GT-7 laser transmitter were inconclusive. Astronauts Borman and Lovell sighted the flashing argon laser beam at two of the three receiver sites (Kauai, Hawaii and White Sands Missile Range, New Mexico on December 11 and 12, 1967, respectively), but no signals were received by either ground station. Although Borman had a very clear view of the argon laser, it was visible for only a few minutes each time. NASA scientists had estimated that 3 to 4 minutes would be required to find and lock onto the receiver station with the hand-held GT-7, so the chances for a successful space to ground link were minimal at best.

None of the remaining Gemini or Apollo crews carried laser-communications gear, although the Apollo lunar missions utilized a ruby laser to range off the moon. Though the initial laser-communicator experiment in space was not totally successful, much was learned from the range performance study of the system and from the design and assembly of both the GT-7 transmitter and the receiver stations. Additionally, the experiment demonstrated for the first time that astronauts could safely view a visible laser beam originating from earth.

Interest in laser-beam space communications remains high. It must be noted, however, that microwave data links pose an impressive challenge to laser systems, particularly since microwave systems are capable of communicating between earth and space vehicles over a distance of millions of kilometers. This topic was discussed in more detail earlier in this chapter.

A particularly interesting space laser communicator has been proposed by the McDonnell Douglas Astronautics Company. This system, which is characterized by small size and

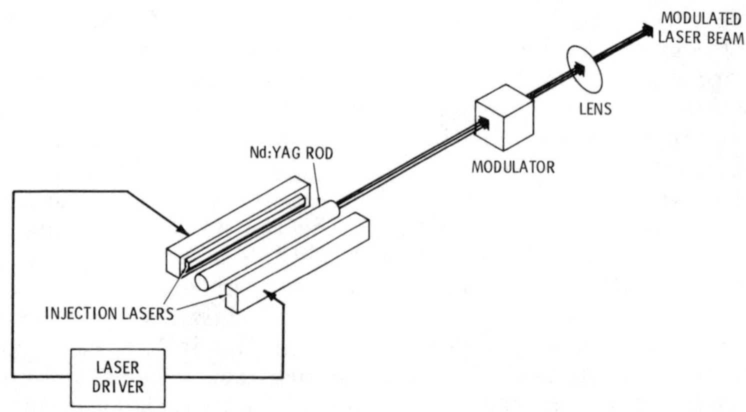

Fig. 6-20. Diode-pumped Nd:YAG laser for space communications.

high efficiency, utilizes a Nd:YAG laser pumped by an array of GaAs injection lasers. The laser diodes are temperature-tuned to emit at the 868.6-nm absorption region of the Nd:YAG rod. Fig. 6-20 is a diagram of the laser and modulator components.

The McDonnel Douglas Astronautics system can utilize pulse-interval modulation (pim), pulse-position modulation (ppm), or pulse-code modulation (pcm) to send up to 10^9 bits of data per second. The overall power efficiency of the system is potentially as high as 10 percent.

McDonnell Douglas Astronautics and Lockheed Missiles and Space have each developed laser transmitters capable of transmitting up to 10^9 bits per second from an earth satellite to aircraft and ground receiver stations. Synchronous satellites might be employed with either of these systems to permit transmissions to a ground station in the United States regardless of the position of the laser-carrying satellite.

Both McDonnell Douglas and Lockheed built their systems in anticipation of Air Force contracts to develop what eventually will be a fully space qualified, gigabit-per-second laser transmitter. The Lockheed system uses a continuous-operation Nd:YAG laser, and the McDonnel Douglas system uses a pulsed Nd:YAG laser. The invisible 1.06-micron wavelength is converted to visible green at 530 nm by a frequency-doubling crystal. This is done to provide a more sensitive wavelength for the photomultiplier receiver. A peak power output of 250 milliwatts in the green is anticipated. Both systems also use 17.8-cm optical antennas to collimate the laser radiation into a beam with a divergence of only 5 microradians. This makes accurate pointing essential, since a 5-microradian beam will illuminate a spot only 300 meters across at a range of 60,000 kilometers.

The high degree of pointing accuracy necessary for proper operation of these two systems is provided by a closed-loop feedback circuit containing the receiver. In operation, the receiver sends an acquisition beam to the laser satellite. Then a quadrant photomultiplier tube on the satellite signals a gimbaled mirror to point toward the receiver site. The satellite activates a pointing laser which illuminates the receiver site. Fine pointing and tracking corrections are automatically accomplished. And finally the modulated laser begins to transmit data.

Within a few years, the Air Force hopes to launch the first of a series of space-qualified laser transmitters into orbit. The construction of these highly sophisticated systems will be directed by the Air Force Space and Missile Organization in Los Angeles.

Meanwhile, NASA, the Air Force, and private companies continue to investigate other methods for communicating between earth and deep space with lasers. Carbon dioxide lasers are receiving considerable attention in this regard because of their very high efficiency and high-quality beam. Direct solar-pumping schemes for solid-state ion lasers are also being considered because of their power-saving advantages. Fig. 6-21

Fig. 6-21. Solar-pumped laser system.

shows a sun-pumped laser and telescope system developed by the Air Force. This system can transmit digital data, voice, and television signals to a conventional laser receiver.

OPTICAL-WAVEGUIDE SYSTEMS

Optical fibers show great promise as waveguides for high-data-rate communication channels. Already several fiber-coupled data links are in operation, and these will be discussed shortly. First, a discussion of optical-fiber advantages and disadvantages is warranted.

Currently, low-loss glass fibers are quite costly, but improvements in materials technology and manufacturing techniques are expected to bring prices down to a more reasonable level soon. Two other problem areas are more permanent—glass fibers cannot transmit electrical power, and fibers require both careful handling and new methods of making connections and splices. One manufacturer has solved the first drawback by simply installing conventional copper conductors in the same cable with several fibers. The fibers carry the data, while the copper conductor supplies electrical power. Handling and splicing fibers are topics receiving a great deal of attention by those involved in fiber manufacture (see Chapter 5).

The advantages of optical fibers outweigh potential disadvantages. The advantages are:

1. Optical fibers have a very high data-rate capacity.
2. Fibers are very small in size.
3. Fibers are potentially very inexpensive since glass is a common substance.
4. Glass is noninductive and therefore is immune to lightning, electromagnetic pulse effects induced by nuclear explosions and inductive cross talk.
5. The very high data rate of a fiber system means that multiplexing can be employed to expand the capacity of a fiber cable over a period of time.
6. Fibers are generally unaffected by moisture and temperature, so cable pressurization is not required.

Several optical-fiber data links are commercially available now, and many others are in a development stage. American Laser Systems, Inc. manufactures two fiber-optic data links. The Model 732 Optical Transmission Set has a bandwidth exceeding 35 MHz and a rise time of less than 10 nanoseconds. The use of a glass-fiber waveguide gives greater than 8-megavolts isolation between the transmitter and the receiver. The 732 has a snr of 50 when a 30-meter optical-fiber cable is employed. A GaAs infrared-emitting diode is used as an optical source, and a silicon avalanche photodetector is used as a detector. Applications for the 732 include signal transmission in or near high-electromagnetic fields, highly secure data links, signal coupling to and from high-tension power lines, and intersystem signal-ground isolation.

The American Laser Systems Model 738 Dielectric Trigger Link supplies up to four laser-trigger pulses with subnanosecond rise times over individual 30-meter lengths of fiber-optic cable. The use of a GaAs injection laser restricts jitter between the four outputs to less than 250 picoseconds. The high radiance of the laser and the exceptional sensitivity of the photodiode permit fiber-optic transmission lengths exceeding 60 meters. The 738 system is used as the trigger source for an 8-megavolt pulse generator with a 10-nanosecond rise time. This pulse generator is designed to subject military aircraft to the simulated electromagnetic pulse effects of a nuclear blast.

Meret, Inc. (1815 24th St., Santa Monica, CA 90404) manufactures a line of miniature optical-fiber transmission sets under the MODAL product designation. The Meret MODAL systems can transmit up to 80 megabits per second over a 30-meter glass-fiber cable. At 80 Mbit/sec, the snr is 10, but the snr is more than 2000 at a 4 Mbit/sec rate. The Meret MODAL fibers are terminated by attaching the fiber ends directly to miniature transmitter and receiver modules only 1.27 cm in diameter.

Commercial fiber-optic data-transmission gear is available from several other firms, and the number of firms can be expected to grow rapidly. Already the military is expressing very strong interest in fiber optics for operational roles. The Navy Electronics Center, for example, has installed a fiber-coupled intercom aboard a 6th Fleet ship. The system utilizes GaAs LEDs and consists of a central switching station and six remote terminals. Gallileo Electro-Optics supplied the 30-meter lengths of low-loss fiber-optic cables used in the system. Each plastic-sheathed cable contains 300 multimode glass fibers.

Recently the Air Force flight-tested an aircraft fitted with a fiber-optic control system. The Hughes-designed system permits two-way data transfer and is immune to lightning and nuclear-pulse effects. It also weighs less than an equivalent wire cable system. Marconi-Elliott Avionics has designed a similar fiber-optic link for the British Royal Aircraft Establishment, and several other firms in the United States and abroad are investigating this new technology.

Missile-data links are a particularly important application for optical-fiber data links. The Boeing Company has proposed such a system to replace the conventional cables used on missiles. Weight reduction would be an important benefit because of the smaller size of the fiber-optic cables and the reduced size of the cable housings. Indeed, the Boeing study found that cable housings could be eliminated by imbeding the fibers directly into the missile skin. A particularly important benefit of fiber data links is immunity to electromagnetic effects. Conventional missile data links are susceptible to lightning, nuclear-pulse effects, and potential differences caused by the very high charge created by the rocket exhaust.

The greatest potential application for fiber-optic data links is the telephone system, and ITT and Bell Laboratories have developed several working transmitters, receivers, and repeaters.

CONCLUSION

The widespread development program in optical communications confirms Alexander Graham Bell's conviction that the photophone was his most important invention. Many researchers agree that operational LED and laser communicators will begin to have an important impact on communications by 1980—exactly 100 years after Bell and Tainter first communicated with one another over modulated beams of reflected sunlight.